THE SCIENCE OF GIS LEADERSHIP
You're Not Just Leading Teams. You're Shaping The Future.

Paperback ISBN: 978-1-947276-88-8
Hardcover ISBN: 978-1-947276-89-5
eBook ISBN: 978-1-947276-87-1

Publishing and Design:

EPIC AUTHOR
PUBLISHING

Ordering Information: Quantity sales. Special discounts are available on quantity purchases by corporations, associations, and others. Orders by U.S. trade bookstores and wholesalers. For details, contact the publisher below:

Contact: 561-601-9871 | support@epic.media | EpicAuthor.com

First Edition

A PRACTICAL FRAMEWORK TO GO FROM TECHNICAL EXPERT TO ENTERPRISE LEADER

THE SCIENCE OF GIS LEADERSHIP

YOU'RE NOT JUST LEADING TEAMS. YOU'RE SHAPING THE FUTURE.

IZABELA MILLER

To my daughters, Tiffany and Gray.

Thank you for loving me despite my imperfections.

Table of Contents

Foreword

By Matt Pietryszyn

BEFORE I EVER STEPPED into the world of GIS, I fell in love with maps, starting as a young boy with a paper roadmap unfolded on the floor, marker in hand, tracing my own imaginary town boundaries and streets. Later, my appreciation of beautiful symbology and the clever logic of geography blossomed. And, after twenty-five years of leading GIS teams in the public sector, I began to understand that the real value of our work is rarely in the maps themselves. It lives in how spatial thinking becomes part of how people work and how decisions are actually made.

That distinction, between making things and making things matter, sits at the heart of *The Science of GIS Leadership*. In this book, Izabela Miller focuses on what is often overlooked, bringing structure to the human side of GIS leadership and shaping it into patterns we can recognize, test, and adapt in our own workplace.

Our tools have matured, data is widely available, and expectations continue to rise, and yet translating that capability into everyday decision-making remains difficult. This book does not chase shortcuts, but instead helps make sense of the quiet, intuitive work we've been doing but usually struggle to describe.

One of the most powerful lessons I have learned is reflected throughout these pages. The job is not to get people into your GIS, but to get GIS into where the people in your organization already work. We have all seen technically excellent maps or applications fail because they asked users to change the way they work to access insight. Real impact emerges when spatial context is embedded into the systems, conversations, and moments where decisions already happen. The real value of GIS appears when people rely on it without having to think about it.

Izabela's interviews with more than fifty GIS leaders and professionals are the backbone of this book. As you read, much of this will feel familiar–the frustrations, the quiet wins, the sense of carrying enterprise expectations without enterprise support, and the moment when being the "map person" is no longer enough. What connects these voices is not a shared platform or organizational model, but a similar way of thinking about impact and responsibility.

This is also where the word "science" earns its place in the title, not because the book is academic, but because it is pattern-based. The impact of sustainable GIS takes shape when its purpose is clear, systems are designed to hold under pressure, and leaders stay attentive to the people doing the work. Skill without self-awareness eventually breaks down under pressure. Vision without systems does not hold. Systems without stewardship lose their integrity over time. When these elements align, GIS shows up where it matters most, at the decision table.

For those early in their careers, this book offers reassurance and direction. You were not hired to make maps. You were hired to help people make better decisions, faster, and with fewer obstacles. For seasoned practitioners, it provides a gentle but necessary challenge to move beyond optimizing for platforms and toward designing workflows that people actually use. For leaders, it encourages a shift toward treating GIS as an enterprise information service, one measured by reliability, trust, and real impact rather than new features.

What Izabela has created here is not a textbook or a theoretical exercise. It is a field guide, written by someone who has lived the work, listened carefully to others who live it too, and taken the time to name what often goes unspoken. This is a book you will return to, not because it tells you what tool to buy next, but because it helps you think more clearly about how GIS creates value and what kind of leadership that demands.

If you are holding this book, it's because you care about the future of GIS. Read it with honesty. Notice what resonates. Pay attention to the moments that feel uncomfortably familiar. Then take those insights back into your own organization and put them to work. The promise of GIS has never been about better maps. It has always been about better decisions, made possible through thoughtful, disciplined, and human leadership.

—Matt Pietryszyn
A spatially adjusted data leader
January 2026

INTRODUCTION

Crossing the Ocean

THIS JOURNEY DID NOT begin at a desk. It began with a decision.

I tucked my Geography degree into a suitcase, left the comfort of home, and crossed an ocean toward a future I could feel but not yet see. My English was imperfect, the software unfamiliar, the names of places new on my tongue—but the love of mapmaking was already in my bones.

I began as an intern and climbed: technician, specialist, analyst. I learned how to solve problems with spatial data, how to make maps that answered questions people didn't know how to ask. And along the way, I kept noticing what others missed: technology sleeping in the corner, potential underused, and the quiet power unlocked when Geographic Information System (GIS) and Information Technology (IT) move with strategy instead of just responding to emergencies.

That was the first identity shift. I realized I didn't just want to contribute; I wanted to lead.

The move from doer to leader surfaced the oldest fears: not enough, not ready, not allowed. I was promoted for being technically strong, not because anyone had invested in my leadership. I knew how to configure systems and build apps; I did not know how to set direction, shape culture, or say no without feeling guilty. And I was not alone.

Over and over, in my own career and in the stories others shared with me, the same pattern appeared—people are promoted into leadership because they are exceptional technical contributors. They receive a new title, bigger responsibilities, and very little preparation for what leadership actually requires. The organization assumes the skills will "just transfer." They don't. So the gap continues to grow.

For years, I felt that gap in my own body. I worked harder. I stayed later. I tried to be the hero. It took loss, reflection, and a lot of uncomfortable honesty to even admit what was happening: I was designing systems for others while quietly running my own life on autopilot.

At some point, I realized I could not keep waiting for someone else to fix it. I had been thinking about writing a book for two years, talking myself out of it with the same story many of you know well: Who am I to do this? Who am I to say anything about leadership when I'm still figuring it out myself?

Then I made a difficult decision: to start before I felt ready.

I left my full-time job, stopped treating the book as a someday project, and started treating it like real work. I began inviting GIS professionals to share their journeys—the wins and the burnout, the pride and the doubt, the promotions that came with no training manual. I opened my calendar and my heart, and so did they.

I spoke with people leading city and county programs, utilities, private companies, state agencies, universities, and consulting teams across the globe. Some had "manager" or "director" in their titles. Many did not, yet they were clearly leaders: building trust in hallways, holding systems together in the background, becoming the person everyone called when the stakes were high.

Patterns emerged quickly.

People never said, "I went into GIS because I wanted a corner office." They said things like, "I wanted GIS to matter enough that I had to step up," or "I didn't plan to lead; I just couldn't stand seeing this potential wasted." Many described being a team of one, or "the only GIS person in the building," carrying the weight of enterprise expectations without enterprise resources.

In parallel, I created a survey for professionals transitioning from technical roles into leadership. Their responses echoed what I was hearing in the interviews:

- They wanted a clear path that didn't require them to become someone they were not.
- They craved mentors, sponsors, and frameworks that were both human and rigorous.
- They were tired of feeling like "the map person" in rooms where real decisions were being made.

This book grew out of those conversations and that tension.

It is not a leadership textbook or a victory lap. It is a field guide written by someone who has sat where you sit, carrying tickets, projects, expectations, and a quiet sense that you are capable of more than "just keeping the lights on."

The science as I have lived it.

The title of this book is **The Science of GIS Leadership** for a reason.

There is a pattern underneath the noise; a structure you can learn, test, and adapt. When I look at my own journey and the journeys of the leaders I interviewed, sustainable impact appears when three forces reinforce one another:

- You **See with Clarity**: who you are, what the work is for, and which decisions GIS must enable.

- You **Scale with Systems**: translating vision into a runnable operating system—roadmaps, request paths, governance, budgets, data practices—that fit your reality.
- You **Serve with Stewardship**: creating conditions where people and systems can perform at a high level without burning out.

Skill without self-awareness fractures under pressure.

Vision without systems fades.

Systems without stewardship decay.

When these three come together, adoption rises, credibility grows, and the value of GIS becomes visible where it matters most—at the decision table.

You will see this structure throughout the book.

Part I: *See with Clarity* focuses on identity, vision, and the North Star that aligns your work with your organization's mission.

Part II – *Scale with Systems* turns that vision into a 90-day roadmap, a longer-term strategy, and practical systems for work, data, budgets, and executive support.

Part III – *Serve with Stewardship* zooms in on your team: how you hire, protect, align, and grow people.

Part IV – *Serve from the Inside Out* turns the mirror fully toward you: your traits, your listening skills, your relationship with conflict, your ego, your support systems, and the shift from scarcity to influence.

There is a reason this framework is built around three moves.

In rhetoric, in storytelling, in change, the human brain gravitates toward threes. We remember beginnings, middles, and ends. We feel the completeness of "past, present, future" or "thoughts, feelings, actions." Three is just enough to create a pattern without becoming a blur. **See. Scale. Serve.** It is simple enough to remember on a hard day and deep enough to hold a whole career.

But this is not about fitting your life into my model. It is about giving you a map sturdy enough to hold your context, your constraints, and your voice.

You are not just building maps.

We are not only mapmakers. We are builders of conditions for better decisions.

We run an enterprise information service—a system of systems that lives alongside finance, Human Resources (HR), operations, and other core functions. It sits in the same architecture diagrams, speaks the same language of risk and value, and is judged by the same standards of reliability, security, and impact.

Leaders I spoke with reminded me of this in different ways.

Some described the moment they stopped calling themselves "the GIS person" and started saying, "I run an information service that supports how we invest, respond, and plan." Others talked about how everything changed once they treated GIS as a real system—with intake, service levels, governance, and metrics—instead of as perpetual emergency response.

They also described the cost of not making that shift: burnout, being stuck in ticket land, invisibility when strategy was discussed, and being last in line for budget and staff.

Many of them were promoted because of their technical excellence. Very few were given the leadership training they needed. They had to design their own.

If that is you, this book is meant to shorten the path.

You won't find a tour of leadership styles here. There aren't any a personality quizzes or a promise that if you just "act like a CEO," everything will fall into place. Instead, you'll find principles, stories, and tools arranged in a sequence that mirrors real-world experiences:

- First, you shift your identity from hero to builder.
- Then, you connect your work to a vision and a roadmap that people can understand.
- Then, you redesign the way work and data move so the plan survives real life.
- Then, you build a sustainable team and culture that can carry that work without burning out.
- And throughout, you cultivate self-leadership with more honesty, courage, and love.

You will hear my story in these pages. You will also hear echoes from the people who trusted me with their own: GIS managers who are teams of one; directors leading multi-disciplinary departments; analysts who lead sideways and upward without a formal title; executives who were once "just the map person" and now sit at tables where major decisions are made.

Where I share their experiences, I do it with respect for their context and, when needed, anonymity. The point is not to build a hall of fame. The point is to show you that you are not alone.

Throughout the book, you'll find principles and practices sturdy enough to hold your voice, your context, and your constraints. We will talk about promotions without training, calendars stacked with urgents, teams stretched thin, grief and burnout, ego and politics, and the pressure of being "the map person" in rooms where the stakes are high and the understanding is low.

This book is for the GIS analyst who keeps getting pulled into strategy conversations without a map for their own leadership.

For the mid-level manager living on the edge of burnout, trying to protect their team from constant fire drills.

For the "team of one" who suspects there is more possible than quiet heroics and reactive work.

For the executive who senses that GIS could be more than a line item, but doesn't yet have language for what that might look like.

As you read, I invite you to bring your whole self: the part of you that loves the work, the part that is tired, the part that still whispers, "Who am I to lead at this level?" You will meet other leaders who have asked the same question.

All of them helped shape this map.

My daughters, Tiffany and Gray, will grow up in a world shaped by the decisions people like you and I make—about infrastructure and climate, housing and health, equity and safety. They don't see my job titles. They see whether I am present when I'm with them. They see whether I stand up for what matters or stay quiet when it's inconvenient. They see whether I treat myself with the same respect I ask them to have for themselves.

When I think about leadership now, it isn't abstract. It sounds like two questions:

1. What kind of leader do I want them to see when they look at me?
2. And what kind of world am I helping build with the decisions I make every day?

How to use this book.

You do not have to read this book in a single sprint. It is not designed that way.

Each chapter ends with a short Leader's Map in Practice section—a few reflection prompts to help you turn ideas into decisions, conversations, and experiments in your real context. You can use them alone, with a mentor or coach, with your team, or with a peer circle of GIS leaders.

You will not implement everything at once. Please don't try.

Instead, treat this as a map you return to at different times in your career:

- When you feel stuck in identity, come back to **Part I**.
- When you need to turn big ideas into reality, revisit **Part II**.
- When your team is tired or fragmented, live in **Part III** for a while.
- When you sense that the biggest bottleneck is inside you, sit with **Part IV**.

My promise is not that this book will make leadership easy. It won't.

My promise is that you will feel less alone, more equipped, and more honest about both your constraints and your possibilities. You will have language for conversations you've been postponing, tools for decisions you've been avoiding, and stories that remind you why you started.

You have already crossed oceans in your own way—leaving old roles, old identities, old versions of yourself behind. This book is an invitation to cross another: from technical expert to enterprise leader, from scarcity to influence, and from reacting to designing.

You don't have to wait until you feel ready.

You get to begin now.

Big love,
—Izabela

Part I:
See with Clarity

BEFORE YOU CAN RUN an enterprise GIS program, you have to answer a quieter question:

What exactly am I here to lead?

Most GIS professionals are promoted because they are excellent doers. They make beautiful maps, fix broken services, and rescue failing projects. But leadership asks for something different: it asks you to move from *hero* to *builder*—from "I can fix it" to "I can design a system where others can win."

Seeing with clarity is about three layers of sight.

First, **identity**: you are not "just the map person." You are running an information service that shapes how your organization sees itself and the world it serves. Until you own that identity, no one else will.

Second, **vision**: not a slogan on a slide, but a picture of a future you can actually describe in plain language—a future where GIS is woven into how decisions are made, not bolted on at the end. Vision that people can see and locate themselves inside.

Third, **honest self-assessment**: understanding your strengths, your gaps, and the limits of what you can do alone. You cannot lead from a fantasy about yourself. You lead best from the truth.

In this part, you'll meet leaders who started as or still are a team of one, analysts who quietly became the person everyone trusted, and mentors who saw the leader before the title existed. You'll be invited to look at who you are now, who you're becoming, and what you want GIS to make possible in your organization.

We start here because everything else in this book—the systems, the team, the culture—will rest on how clearly you can answer three questions:

Who am I as a leader?

What is this work for?

Where are we actually trying to go?

Once you can answer these questions with honesty, you're ready for the next challenge: translating that clarity into actionable steps.

CHAPTER 1:

The Identity Shift: From Mapmaker to Information Leader.

IF YOU WORK IN GIS long enough, people will give you a nickname.

"The map person."

"The GIS gal."

"The Esri guy."

"The one who makes the magic happen."

At first, it can feel flattering. You're the one they call when something needs to look better, make sense, or get fixed by Friday. You swoop in with a map, a dashboard, a quick analysis, and people say, "I don't know how you did that, but thank you."

For a long time, that was enough for me.

I loved solving problems. I loved watching someone's shoulders drop when a confusing mess turned into a clear picture. I loved turning raw data into something that made sense on the screen.

But as my career unfolded—and as I listened to dozens of GIS leaders around the world—I began to see a pattern:

Being "the map person" is a beautiful place to start.

It is a terrible place to get stuck.

Because at some point, the problems you care about most can't be solved with another map alone. They require decisions, priorities, funding, and coordination. They require people in rooms you're not yet invited into to see GIS as more than a service they occasionally tap for visuals.

They require you to become something more than the person who makes the map.

They require you to see yourself—and behave—as an information leader.

This chapter is about that shift.

When your job title changes but your identity doesn't.

Most leadership journeys in GIS don't start with a grand announcement. They start with... more work.

You're the one who understands the database, so you're asked to "keep an eye on" it.

You're the only person who knows the web stack, so now you're "leading" app development.

Someone needs to speak in a meeting, and your manager says, "Can you walk them through what we're doing?"

Suddenly, without much ceremony, you're the one connecting pieces, explaining patterns, and calming people down when technology misbehaves.

People start looking to you not just to produce work, but to guide it.

But here's the catch: your calendar often doesn't change.

You still carry the same tickets, the same production work, the same "can you just..." calls. On top of that, you're now expected to think strategically, meet with executives, mentor juniors, and represent "GIS" in bigger conversations.

Your job description has quietly doubled. Your sense of identity is still catching up.

When I talk to GIS managers and emerging leaders, this is where they often feel the most tension. Their badge says "Manager" or "Analyst," but inside, they still feel like an overworked GIS specialist who happens to have more meetings.

One leader told me, "I'm supposed to be shaping strategy, but my day is still filled with doing the work. I don't feel like a leader. I feel like an exhausted technician with a nicer title."

I understand that feeling intimately because I lived it.

From production hero to enterprise builder.

If you've spent years being rewarded for "getting things done," it's natural to define your value by output.

You become the person who says yes. The one who can be counted on. The one who jumps in, saves the day, and pulls the all-nighter.

In the early stages of a career, this can be incredibly useful. You learn fast. You touch many systems. You build trust.

But as you move into leadership, the strengths that helped you rise can quietly become the very things that hold you—and your team—back.

You can't be the hero and build a healthy system at the same time.

A GIS manager I interviewed put it this way: "I realized I was the bottleneck in a system I had designed. I was proud of everything I'd built, but I hadn't built it to run without me."

Another, **Jim Alberque**, a GIS and Engagement Technology manager at the City of Raleigh, talked about moving from managing one person to leading a much larger team. The turning point for him wasn't a new technology. It was realizing he had to stop measuring his day by what he personally produced and start measuring it by what the team was able to accomplish together.

Adam Carnow, the Public Works Industry Marketing Specialist at Esri and keynote speaker, talked about how often there is "just one person" trying to move GIS forward in a department—the only one who can see the potential. That can feel heroic for a while. Over time, it becomes heavy. When you only see yourself as "the lone champion," you carry everything alone, and it becomes hard to ask for the support and structure you actually need.

One leader told me he fell in love with GIS twice. The first time was all about maps—colors, symbology, and the craft of making something beautiful. The second time was when he realized the real work was getting GIS into the hands of people who would never call themselves GIS users. He started asking, "How do I make this easy for a planner who dislikes technology? For a manager who only has five minutes? For a field worker on a bad connection?" That question changed everything. He stopped thinking of himself as a mapmaker and started seeing himself as a designer of access—someone whose job was to make location-based information usable for the people doing the work.

The shift from production hero to enterprise builder is not about working less or caring less. It's about redefining what "good work" looks like in your role.

A production hero asks, "What did I finish today?"

An enterprise builder asks, "What system did I improve today so that more work can happen reliably, even when I'm not in the room?"

This is where *See with Clarity* really begins. Before you build a single roadmap or governance process, you have to decide who you are in this story: the exhausted hero who saves the day one ticket at a time, or the builder who designs conditions where better decisions are possible every day, with or without you.

This is not easy. Your brain will miss the quick wins and instant gratitude of fixing things yourself. Your ego will miss being the expert.

But this is where leadership begins.

Teams of one are still leaders.

You might be listening to this and thinking, "That's nice, but I don't have a team. I *am* the GIS department."

I want to say this very clearly: the identity shift matters just as much—maybe even more—when you're a team of one.

Several of the leaders I spoke with started exactly there: one person in a city department who happened to know GIS, one analyst at a utility who started building better maps and slowly became the person people trusted, one specialist in a small community who wore every hat—data, apps, training, support.

Alexa Venezia, a GIS Specialist for San Antonio, for example, spoke about this tension from the perspective of a GIS specialist in local government. She didn't have a staff reporting to her, but she had a constant stream of requests, competing priorities, and the nagging sense that "being helpful" and "building something sustainable" were not the same thing.

People like Alexa didn't wait for a management title to begin leading. They started acting like stewards of an information service, even while their email signatures still said "analyst" or "specialist." They learned to ask better questions. They began to push for standards, not just quick fixes, and educated their colleagues on what was realistic.

They didn't have authority on the organizational chart. They built authority through consistency, clarity, and care.

One leader told me, "I realized I was the only one in the room who understood how these systems connected. That came with responsibility. Even if my email signature didn't say 'manager,' people were making decisions based on what I said."

If you are a team of one, this book is for you, too.

You may not be able to change your org chart overnight, but you can change the story you tell yourself about what your work is for. You can move from "I make maps" to "I help this organization see itself clearly enough to make better decisions."

If you're a team of one, you may not be able to hire people or change your job title tomorrow, but you can still shift your identity from "I make whatever anyone asks for" to "I steward an information service that supports this organization's mission."

That shift alone will change the conversations you have.

Your calendar tells the truth.

We can talk about identity and mindset all day, but one of the most honest mirrors of your leadership is your calendar.

If your schedule is filled with other people's urgent requests, if you have no focused time set aside to work on important tasks, if every week feels like a scramble to address demands, then no matter how good your intentions are, your impact will be limited.

Part of becoming a leader is learning to treat your time as a strategic resource, not just a container for everyone else's priorities.

In a conversation with **Wendy Baldin**, a Director of Innovation, Technology, and Geospatial Solutions in Ontario, Canada, we talked about the importance of protecting time on her calendar for thinking, planning, and growing.

She said, "Being able to take that time and guarding it as *gold*—that is valuable time for my career, not only to be able to help myself professionally, but also to be able to help the business."

That sentence captures the identity shift beautifully.

You are not being selfish when you block time to think, learn, plan, or design better systems. You are doing your job as a leader and increasing your capacity to serve.

Leaders who refuse to guard that time eventually lead from depletion. They make reactive decisions and start resenting the very people they're trying to help. Their teams pick up on the hurry and frustration, and it becomes the culture.

The identity shift is not just a feeling. It's visible in what you allow on your calendar, what you say yes to, and what you protect.

The story you tell yourself about your role.

Underneath all of this is a story.

Most of the GIS professionals I've interviewed carry a story about their role—sometimes spoken, often unspoken:

"I'm just here to support."

"I'm the one who fixes things when nobody else can."

"My job is to make whatever map they ask for."

"I don't really have a say in strategy."

If you listen closely, you can hear the limits in those sentences.

During one conversation, a GIS manager admitted, "I've been acting like my job is to keep everyone happy, not to lead." That realization stung. But it was also freeing, because it gave them permission to change that story.

What if your story sounded more like this?

"My job is to help this organization make better decisions with the power of location."

"My job is to steward the GIS program—its people, its data, its systems—so it can serve for the long-term."

"My job is to translate between the language of maps and the language of executives."

Same skills. Same person. Different identity.

When you start thinking of yourself as an information leader, you ask different questions in meetings. You push for clarity on priorities. You take responsibility for educating others about what GIS can and cannot do. You stop apologizing for needing standards, time, or staff to deliver on promises.

The organization may not have caught up yet. Titles and structures move slowly. But you don't have to wait for the org chart to validate your identity.

You can begin leading from that place now.

You don't need permission to start.

Here is the most important thing I want you to hear from this chapter:

No one is going to email you one day and say, "**Congratulations, you are now officially an information leader. Please start behaving accordingly**."

The shift will happen quietly, in the way that you decide to see your work.

It will happen when you stop measuring your value only by how many tickets you close and start measuring it by how much clarity and stability you create for others.

It will happen when you allow yourself to ask, "What do I see that others don't yet?" and "What am I willing to stand for, even if it's uncomfortable?"

It will happen when you choose to build systems that work without you as the hero.

This doesn't mean abandoning the craft you love; your technical skills are part of your power, giving you credibility and empathy. You know what it takes to actually build the things people request.

But they are the foundation, not the ceiling.

As you move through the rest of this book, we'll talk about roadmaps and governance, team building and conflict, influence and stewardship. All of that will be more powerful if you start here:

You are not just "the map person."

You are not just the person they call when something breaks.

You are part of shaping how your organization sees the world and makes decisions in it.

You are simply shaping the future.

You don't need to be perfect. You don't need to have every answer. You don't need anyone's permission to begin.

You only need to decide today to start seeing yourself as the leader you are becoming—and then take one small step, and another, and another in alignment with that decision.

The maps you make will still matter. They always will.

But the conditions you build—for your team, your data, your systems, and yourself—will be the real legacy of your leadership.

Leader's Map in Practice - Identity Shift

Use these as a short pause before you move on.

1. **The story you're living now:**

   In one honest sentence, tell the story you're living now. How do you currently view your role?

   Then write one sentence for the role you're growing into.

   - "Right now, I see myself as..."
   - "I want to start seeing myself as..."

2. **Your calendar as a mirror:**

   Look at last week. Where did you act like a production hero, and where did you act like a builder of a system?

   - Name one small change you can make in your calendar next week that's more aligned with being an information leader (for example: protecting one focus block or saying no to one "urgent" that doesn't matter).

3. **Team of one is still a leader:**

   Whether you have a team or not, finish this sentence:

   - "If I treated GIS as an enterprise information service I steward—not just work I do—I would start by..."

Write your answers in a notebook, a document, or the notes app on your phone. The point isn't perfection; it's noticing where you are and where you want to go.

CHAPTER 2:

A Vision People Can Actually See

MOST GIS "VISIONS" ARE born in PowerPoint.

You've probably seen them.

"Innovative, data-driven, geospatially enabled organization."

"Single source of truth."

"Empowered decision-makers through location intelligence."

They sound impressive, but they also blur together. After the slide deck closes, no one can really tell you what any of it would look like on a Tuesday morning.

The irony is that GIS people are experts at making the invisible visible.

We spend our days turning messy information into something someone can see, touch, and act on. But when it comes to our own vision, we often settle for vague phrases that could belong to any tool, in any organization, in any decade.

If you want executives, peers, and your own team to care about your vision, it has to pass a simple test: someone who is not you should be able to picture their world changing because of it.

In Chapter 1, we started with **See with Clarity** on the inside: the shift from "map person" or "production hero" to information leader and builder. This chapter stays in that same movement but turns your attention outward. Now that you're starting to see yourself differently, you need a vision for GIS that is just as honest and concrete—a vision other people can actually picture and repeat when you're not in the room.

When I first became a GIS manager, it wasn't because I was the most experienced person in the room, but because I had a vision.

I could see what GIS could do for the organization in a way that felt almost physical. I saw how underused the software was and the gaps between what leaders were worrying about—budgets, regulations, growth, risk—and what our maps and systems could actually help them understand. I could picture dashboards that didn't exist yet, conversations that could be different, and decisions that could be better informed.

That vision is what helped me step into leadership.

But very quickly, I ran into a wall: the picture in my head was clear to me but fuzzy to everyone else.

I could talk about possibilities, but I didn't yet know how to turn that into something people could see, test, and believe. I didn't have a way to describe our current reality in terms executives could act on. I couldn't show them, in concrete terms, what would be different if we invested in GIS as an information system instead of as a side tool.

The problem wasn't that I lacked vision.

The problem was that my vision lived mostly in my language, not theirs.

I've learned since—in my own work and from the interviews for this book—that many GIS leaders are in the same place: the vision is there; it's just not yet visible.

Lofty words vs. a shared North Star.

Most organizations already have vision statements.

They mention words like innovation, resilience, sustainability, excellence, community, and safety. They sound impressive on posters and websites, but they rarely change anyone's calendar.

The same thing can happen with GIS.

I've seen plenty of GIS "visions" that sound like this:

- "To be a world-class provider of geospatial solutions."
- "To leverage GIS to support decision-making across the organization."
- "To be the leader in spatial data and mapping services."

There is nothing wrong with those words. The problem is that they don't answer the questions people quietly care about:

- What problems will this actually help us solve?
- Which decisions will become less risky, less political, or less painful?
- How will my day-to-day work improve if we take this seriously?

A real vision is not a slogan. It is a shared North Star—a picture of a future state that people can describe in their own words and recognize when they see it starting to happen.

When I asked leaders in the interviews what their vision looked like, their most powerful answers were not about technology. They sounded more like this:

- "I want any department head to be able to answer their top three questions with confidence in one place, instead of hunting through five systems."
- "I want council meetings where people don't argue about whose map is right."

- "I want field staff to trust the data in their hands the way they trust their tools—not as a nice-to-have, but as something they rely on every day."

You can feel the difference. These are things you could argue for. They are specific enough that people can tell if you are getting closer.

That is what we are aiming for.

GIS vision cannot live in a separate universe.

One of the biggest mistakes I see GIS leaders make—and that I made early on—is treating the GIS vision as something separate from the organization's.

We talk about "our" roadmap, "our" architecture, "our" adoption challenges, as if GIS is its own little planet orbiting the business at a distance.

But GIS is not special in that way. GIS is an information system.

Just like finance, HR, Customer Relationship Management (CRM), and asset management, it exists to support the organization's mission and goals. If your city's vision is about safe, livable neighborhoods, then your GIS vision needs to show how location intelligence will make neighborhoods safer and more livable. If your utility's strategy is about reliability and regulatory compliance, your GIS vision needs to show how location intelligence will reduce outages and help you stay ahead of risk.

A GIS vision that is disconnected from the business vision will always feel like an add-on—something to cut when budgets are tight or attention shifts.

A GIS vision that is nested inside the business vision, on the other hand, becomes harder to ignore.

When I began to understand this, my own language had to change. Instead of saying, "We want to modernize GIS," I started saying things like:

- "We want to give finance, engineering, and operations a shared view of where we are taking risk, so they can plan together instead of fighting over spreadsheets."
- "We want to reduce the time it takes to answer the council's top five recurring questions from days to minutes."
- "We want our community-facing apps to match the level of transparency our residents say they expect."

Those sentences didn't mention software at all; they talked about outcomes.

The leaders I interviewed who had gained the most traction with their vision had all made a similar move. They stopped insisting that people care about GIS as such. Instead, they made the vision about the things their organization was already committed to: safety, equity, growth, cost control, and service quality.

GIS became the how, not the what.

Three questions every clear vision answers.

You don't need a perfect vision statement to move forward, but you do need to answer three questions in a way that people outside GIS can understand.

First: What is this work for?

Not in general—in your organization, at this time.

Is it to reduce risk and surprises, support smarter capital investments, provide better service to residents or customers, make compliance less painful, or create a shared picture that reduces conflict?

Be honest. Pick two or three that genuinely matter to the executives right now. This will change over time, but you need a starting point.

Second: Which decisions must GIS support?

I intentionally use the word *must* here.

If GIS disappeared tomorrow, what decisions would become blind guesses again? If GIS were operating at its full potential, which decisions would feel noticeably different?

Think about annual budget cycles, where to build or repair first, how to respond to storms or emergencies, where to invest in infrastructure, and how to communicate with the public. Those are the places where your vision has to land.

Third: What will be different in three years if this vision becomes real?

Three years is far enough out that you can dream, but close enough that it has to be grounded in reality.

You might say things like:

- "In three years, every major capital decision starts from a shared geospatial picture that finance, engineering, and leadership trust."
- "In three years, our data governance is strong enough that we don't argue about whose numbers are right in front of the council."
- "In three years, field staff are co-designers of our GIS tools, not just users or people we push technology onto."

If you can answer those three questions in plain language, you are no longer waving your hands—you are drawing a picture.

The hard part: making it a shared picture.

Naming your vision once is easy. Turning it into something people share is not.

This is the part that doesn't show up on strategy slides.

It looks like you walking into conversations where people are busy, tired, and skeptical, and saying, "Can I show you what I think we could do here?" It looks like you noticing when someone's attention is gone and saying, "This isn't the right moment. Let's grab lunch and talk when you have more space."

It is hard work. It takes time and dedication, grit, and a lot of underappreciation. People might brush you off because they are overloaded or distracted. You have to be conscious of that. You learn to read the room, to feel whether there is even a small opening for the conversation you want to have.

Sometimes the right move is to plant a tiny seed—a story, a question, a sketch—and then leave it alone for a bit. Sometimes it is to ask for a dedicated meeting away from the noise, where you can walk through a scenario together and ask, "What if this looked different?"

Your job is not to force your vision on people. Your job is to keep the soil moist and the climate just right for the idea to grow.

You water it with small wins, with consistent language, with respect for other people's priorities. You stay the course, even when it feels slow. Nothing that matters is easy. Vision is no exception.

What I heard in the interviews.

When I listened back to the interviews with GIS leaders, I noticed that almost all of them had some kind of vision moment—a point when they could suddenly see a different future for their program.

Sometimes it came from a crisis, when a storm, a pandemic, or a major failure made it painfully obvious that doing things the old way was not sustainable. Sometimes it came from exposure—seeing what another organization had done and thinking, "We could do something like that, but in our way." Sometimes it came from frustration—watching the same problems repeat and deciding, "I have to stop treating this as normal."

But the leaders who made progress had one thing in common: they stopped keeping that vision a secret.

They started talking about it in meetings that weren't specifically about GIS. They started using language that made sense to operations, finance, HR, and executives. They connected their vision to the problems other people already cared about.

One leader described it this way: "At first, I felt like I was dragging my vision around the building in a backpack. I knew it mattered, but no one else had asked for it. When I finally started sharing it—carefully, in their language—I realized some people had been waiting for someone to say exactly that."

Your situation will be different. Your constraints are real. You may not have immediate support. You may have been burned before for "being too ambitious."

Even so, the pattern holds: vision grows when you stop treating it as a private hope and start treating it as a conversation.

Testing your vision in plain language.

A simple way to test whether your vision is ready is to run a small experiment.

Imagine you are explaining your GIS vision to three different people:

- A frontline staff member who uses your maps in the field.
- A department head who cares mostly about budgets and risk.
- A community member who will never log into your systems but is affected by the decisions they inform.

Can you describe what you are trying to build in two or three sentences that each of them could repeat later without losing the meaning?

If your explanation only works for other GIS people, it is not ready yet. That is not a failure. It is an invitation to refine.

You do not have to get this perfect before you move on. Vision is something you will keep revisiting. It will sharpen as you learn more about your organization's politics, constraints, and possibilities.

What matters now is that you are willing to name a direction that is bigger than "more maps" and honest about what GIS is actually for.

In the next chapters, we will talk about mentors, influence, and the skills and limitations you bring to this work. Then we will start turning this vision into a 90-day roadmap and, later, a longer-term strategy.

For now, your job is to bring your vision out of your head and into the light.

Leader's Map in Practice - Chapter 2

Pick one or two of these. Write real sentences, not bullets. You are allowed to be messy on the first draft.

1. In your own words, what is GIS for in your organization right now? Not what it could be someday—what you believe it is meant to serve today.
2. Name two or three decisions that GIS must make better if your vision is going to matter. What would be different about those decisions if your vision came true?
3. Draft a short, plain-language explanation of your GIS vision that you could share with a non-GIS colleague. Who will you test it with in the next two weeks, and what will you listen for in their reaction?

In the next chapter, we will look at the people and forces that help you carry this vision—influence, mentors, and honest assessments of your own skills and limits—so you are not trying to do it all alone.

CHAPTER 3:

Influence, Mentors, and Understanding Your Skills and Limitations

A CLEAR VISION IS POWERFUL, but it also weighs a lot if you try to carry it alone.

Once you start seeing yourself as more than "the map person" and can describe what GIS is really for in your organization, the next truth arrives:

You cannot drag this future into existence by yourself. You have to **move people**, not just data.

That is where influence, mentors, and honest self-awareness come in.

This chapter addresses three questions:

1. How does your vision actually travel through your organization?
2. Who helps you grow into the leader you are trying to become?
3. What are you genuinely good at, and where do you need help or growth?

These are not soft side issues. They are the difference between a vision that lives and a vision that dies in your notebook.

The myth of "if I just do good work, they'll notice."

Most of us were raised on a simple promise: if you work hard and produce good work, people will notice and reward you.

In technical roles, that sometimes feels true. You fix a broken workflow. You build a beautiful map. Someone says thank you. Your reputation grows.

But at the leadership level, the pattern changes.

Good work is necessary, but it is **not sufficient**. At that level, people aren't just asking, "Is the work good?"

They are asking:

1. *Does this help me solve the problems I care about?*
2. *Do I understand what it is?*
3. *Do I trust this person enough to attach my name to their ideas?*

That's influence.

Influence is not manipulation. It's not tricking people into your agenda. It's the skill of making it easier for others to say yes to the right things.

In my GIS leadership training, I call this an **influence campaign**: a series of small, purposeful conversations and actions that move your vision from your head into other people's mouths and calendars.

When I started thinking about influence this way, I realized how passive I had been. I was waiting to be "discovered" instead of designing how my vision would travel.

Many of the leaders I interviewed recognized themselves in that pattern. They told stories like:

- "I thought if I just made the best possible map, the value would be obvious."
- "I kept sending long emails explaining our capabilities and wondered why no one responded."
- "I assumed executives would ask about GIS if they cared about it. They didn't."

They weren't failing. They were operating on the wrong myth.

Influence as a map, not a mystery.

The first step in building influence is to stop treating it like magic.

Influence is not "some people have it, and some don't." Influence is a **map** you can learn to read.

Every organization has people who shape decisions, formally and informally. Some sit in C-suite or director roles, while others are long-tenured staff everyone trusts or project managers who quietly decide what moves and what stalls.

Your job is to notice who actually shapes the decisions your vision needs, who they listen to, and how information really flows.

When I coach GIS leaders, I often ask them to draw a simple diagram: the decision they care about in the middle (budget, headcount, a new system, a critical project), then circles of people around it.

Who signs off? Who advises? Who can quietly veto with a raised eyebrow? Who will be most affected but rarely invited into the room?

It's rarely a neat org chart. It's more like a messy spiderweb.

Once you see that web, your influence work becomes more intentional. You stop "spraying" information in all directions and start tailoring conversations. You stop waiting for a single big presentation and start building trust in smaller moments. You stop guessing what matters to people and start asking.

One leader told me that the real shift in her influence came when she stopped trying to "sell GIS" and started asking one simple question in meetings:

"What is the hardest decision on your plate right now, and what makes it hard to decide?"

She listened. Only then did she ask, "Would it help to see that decision on a map or to have a shared picture of the data behind it?"

Influence begins with listening, not talking.

Mentors, champions, and the people who see you before you see yourself.

Very few people make the identity shift from mapmaker to information leader alone.

I didn't.

Early in my career, a mentor named **Trevor Hebditch** quietly began treating me like a leader long before I felt like one. He asked me to run meetings I didn't feel ready for and pushed me to present to people two levels above me. When I protested that I wasn't qualified, he would say some version of:

"You'll figure it out. I trust you."

At the time, I thought he was just being kind. Looking back, I see what he was really doing: inviting me into an identity I hadn't claimed yet. He was lending me his belief until I could build my own.

Years later, after a season of deep personal loss, I began working with a coach named **Robert Ellis**, the founder of *Coaching from Essence*. On paper, I was a leader, with titles, responsibilities, and recurring meetings. Inside, I was still wrestling with the same old questions: *Am I enough? Am I allowed to want more? Who am I if I stop proving myself through constant overwork?*

One day, after listening to me list all the external judges of my worth—titles, budgets, other people's opinions—Robert asked:

"Who decides how much you are worth?"

I started to answer with all those external markers again. He shook his head.

"No. You do. You decide. And then you build a life and a career that honors that decision."

That landed like a tectonic shift. It forced me to see how much of my leadership was driven by fear and scarcity. It also gave me a new kind of responsibility: if I decide I am worth more, I have to act like it.

More recently, preparing to speak at the GIS Leadership Summit, I worked with another coach, Jason Goldberg. That conversation wasn't about maps or technology at all. It was about **what story I wanted to tell from the stage** and **where my ego might get in the way**. It was about deciding whether I wanted to impress people or serve them.

Trevor, Robert, and JG played different roles at different times, but the pattern is the same:

Mentors and coaches don't just teach skills. They help you see yourself more clearly—both your potential and your patterns.

The leaders I interviewed had their own versions of this.

Some had a boss who pulled them aside and said, "You are thinking at a higher level than your title. Let's talk about what's next." Others had a peer who challenged them, "You keep complaining about this gap. When are you going to step into it?" Some found mentors outside of GIS entirely—in leadership programs, community work, or family—who modeled a different way of leading.

If you don't have those people yet, it doesn't mean you are doomed. It means one part of your influence campaign is to **go looking for them**.

You do not have to announce, "Will you be my mentor?" You can start with a small ask:

- "Can I get your perspective on something I'm wrestling with?"
- "Can I share a draft of my GIS strategy and get your reactions?"
- "When you were at my stage, what do you wish someone had told you?"

Mentoring is not a one-way ladder. As you grow, you will also become that person for others—but we'll talk more about that later in the book. For now, it's enough to admit: you were not meant to do this alone.

Seeing your skills and limitations clearly.

Influence and mentoring are powerful, but they're not magic either. They don't erase the reality of your skills and limitations; rather, they help you deal with them honestly.

In the **From Technical to Leadership Journey** survey, many people talked about being promoted because they were good at the technical work, and then being left to figure out leadership on their own. They described feeling like they had two jobs: continuing to be the technical expert, and knowing how to manage people, budgets, politics, and strategy.

No one had helped them distinguish between skills they already had and could lean on, skills they could reasonably grow, or limitations they needed to evaluate, rather than endlessly fight.

Without that clarity, it is easy to fall into two traps:

You either feel like a fraud ("I'm bad at everything that matters") or you cling to being the expert and resist delegating ("No one else can do this as well as I can").

Neither is true, and both will burn you out.

A healthier approach is to treat your capabilities like a **map** instead of a judgment.

You might ask yourself—what kinds of work give me energy and come naturally; what types of work drain me but are still crucial in this role; what kinds of work am I genuinely not suited for, and need to partner or hire around?

Maybe you are strong at seeing patterns in complexity but weaker at detailed follow-through. Maybe you are great with one-on-one conversations but freeze in big rooms. Maybe you are brilliant at designing systems but need help telling the story to non-technical audiences.

None of those things disqualifies you from leadership.

They just mean you need to **design your leadership** around reality.

In the interviews, I heard leaders talk about moments when they finally admitted a limitation out loud—to themselves, to a mentor, or to their team—not in a self-deprecating way, but in a simple, grounded way:

- "I can see the big picture and get us moving. I need someone who loves details to partner with me on execution."
- "I'm good at technical coaching. I'm still learning performance coaching. I'm going to ask for help with that."
- "I can carry this much. Beyond that, the quality suffers. Let's talk about priorities."

Ironically, those admissions often **increased** trust instead of reducing it. People already knew, consciously or not, where the leader struggled, and hearing it named opened up space for more realistic plans.

Seeing your skills and limitations clearly is not about shrinking your vision; it's about dropping the exhausting performance of pretending you're perfect, so you can spend your energy on the work that actually matters.

Apply for the right problems, not just the right title.

Understanding your skills and limits also changes where you choose to focus.

Many of us were taught to chase titles, pay scales, or "the next logical step" in our careers. But the leaders I spoke with who found the most meaning and momentum didn't talk about titles first. They talked about **problems**.

They asked:

- "What kind of problems do I want to spend my life working on?"
- "Where does my way of seeing the world actually help?"
- "What conditions do I need to do my best work?"

I've said to many GIS professionals—and to myself—that applying for roles is not about proving you can do everything in the job description. It's about deciding whether the problems that role is responsible for are the ones you want to take on, with the support and constraints that actually exist.

Sometimes the bravest thing you can do is **not** apply for a job that looks impressive but would bury your strengths and amplify your weaknesses in unsustainable ways.

Sometimes the bravest thing is to **ask** for changes in your current role—more strategic time, clearer expectations, some leadership training—because you see that, without those, your impact will plateau no matter how hard you work.

Influence here isn't just outward; it is also inward: how you influence yourself, your decisions, and the story you're living.

Influence that fits who you really are.

One more thing before we move on: influence does not require you to become someone else.

You do not have to turn into the loudest voice in the room. You do not have to start using buzzwords that make your skin crawl. You do not have to pretend to be comfortable with things that make you deeply uneasy.

Some of the most effective GIS leaders I met were quiet, thoughtful people. They influenced by asking sharp questions, by noticing patterns, and by building deep trust one conversation at a time.

Others were more visible—speaking at conferences, presenting to councils, writing articles—but even then, their power came from authenticity, not performance.

You can feel the difference when someone is there to serve versus to be admired.

The point of understanding your skills, limitations, mentors, and influence map is not to reshape yourself into some generic leadership mold.

It is to **amplify your authentic voice** in a way that serves the work and the people who depend on it.

In the next part of this book, we'll start translating your identity and vision into more concrete structures: 90-day roadmaps, strategies, systems for work, and data.

Before we go there, I want you to carry this with you:

You are allowed to ask for help. You are allowed to grow. You are allowed to build a version of leadership that fits who you actually are.

You don't have to get there all at once. You only have to keep taking one step at a time.

Leader's Map in Practice - Chapter 3

Choose one or two prompts. Be honest and specific; think in terms of names and situations, not abstractions.

1. **Influence map:** Think of one decision you care about (budget, a project, a role). Who actually shapes it—formally and informally? Sketch your influence map: who signs, who advises, who quietly vetoes, who is most affected. Where is one relationship you could start nurturing this month?

2. **Mentors and support:** Who has seen a bigger version of you than you currently see? What is one conversation you could initiate with them (or with a new potential mentor or coach) about the leader you are becoming?

3. **Skills and limits:** List three things you are genuinely good at in your current role, and three that drain you or feel like weak spots. For one of the weak spots, decide: will you invest in growing this skill, or design around it with help? What is a small action you can take in the next 30 days to follow through?

In the next chapter, we will move from people to plans: how to turn your vision into a **90-day roadmap** that reflects your reality, respects your limits, and builds trust in your emerging identity as an information leader.

Part II:
Scale with Systems

VISION WITHOUT A SYSTEM is just a wish.

Once you can see who you are and what the work is for, the next challenge is painfully practical:

How does this actually run, week after week, without burning everyone out?

Many GIS leaders stay stuck in a loop of big ideas and constant firefighting. They have a sense of what's possible, but their days are filled with urgents, interruptions, and half-finished projects. The gap isn't intelligence; it's **infrastructure**—not only servers and software, but the leadership infrastructure that turns intent into repeatable practice.

Scale with Systems is about building a lightweight **operating system for your GIS program**:

- A **90-day roadmap** that connects your vision to concrete work.
- A way to grow that into a **longer-term strategy** that survives budget cycles and leadership changes.
- Language and stories that **promote the value of GIS** as an information system, not an isolated tool.
- Relationships and habits that **earn and sustain executive support**, so you aren't lobbying for funding from the hallway.
- Simple, humane structures for **request management and governance**, so work doesn't just arrive via whoever yells the loudest.

You'll see how leaders use maturity models and strategy frameworks without becoming slaves to them. You'll hear how they shifted from "we'll do whatever comes in" to "here's the work we've committed to, here's why it matters, and here's how new requests fit in."

This part will not hand you a rigid template. Instead, it will offer clear principles and practical examples you can adapt: structure to keep you out of chaos, flexibility to respond to reality, and honesty to say "not now" without burning bridges.

Vision makes you compelling.

Systems make you dependable.

Together, they make you credible.

And yet even the best systems are carried by people. The next part is about those people and the conditions they need to do their best work.

CHAPTER 4:

From Idea to 90-Day Roadmap

BY NOW, YOUR WORLD has shifted.

You're no longer just "the map person." You've started to see yourself as an information leader. You can describe, in plain language, what GIS is for in your organization and which decisions it must make better. You're noticing who shapes those decisions, who mentors you, and where your own strengths and limits lie.

Now comes the question that separates dream from design:

What exactly are you going to move in the next 90 days?

Not someday. Not "in the five-year plan." Not "when things calm down."

The next three months.

Why 90 days is the right unit.

Organizations love big horizons: five-year plans, long-term visions, multi-year roadmaps. Those have their place. But most real change in a GIS program happens in much smaller windows—in focused efforts that last long enough to matter and short enough to survive contact with reality.

When I first tried to turn vision into action, my plans were far too big.

I wanted a modern GIS environment, stronger governance, cross-department dashboards, better data quality, and a clearer story for executives—all at once. I could see the end state so vividly that everything felt urgent.

So I did what many of us do: I created a huge roadmap. It had many colors, many arrows, and many promises. On paper, it looked serious. In reality, it was impossible to run.

Every week, new "urgents" appeared. Requests came in sideways. Projects expanded. The roadmap quickly turned into a guilt document—a reminder of what we were not doing.

It wasn't that the ideas were wrong, but rather the **timeframe**.

Ninety days is long enough to move something real and short enough that reality can't hide. It forces you to be honest about your capacity, the politics, the data and systems, and your own habits.

In the interviews, when I asked leaders what they would focus on in their first 90 days in a new leadership role, the answers followed a pattern:

- Listen more than you talk.
- Stabilize what's fragile.
- Ship one visible win that matters to others, not just to GIS.
- Use that win to build trust for the next step.

A good 90-day roadmap weaves those together. It is not a wish list. It is a small, testable bet on the future you described in your vision.

Start with one honest outcome.

The mistake many of us make is starting with tasks.

We open a document or a project board and start listing things to do: upgrade this server, clean up that data, build this app, meet with that department. Within an hour, the list is overwhelming. No one could finish it in a year, let alone three months.

Instead, start with one honest outcome.

Ask yourself:

If, ninety days from now, one thing had clearly improved in how GIS serves this organization, what would it be?

Not "everything will be better." Not "we will be modern." One thing.

Maybe it is the system that stops falling over once a week. Maybe we stop arguing about which dataset is right for a critical decision, or maybe one cross-department project moves from idea to reality. Maybe we have a shared, documented intake path instead of chaos. Maybe I finally have two mornings a week blocked for strategic work, and that time is respected.

This outcome doesn't have to impress the masses. It has to matter to the people who live with the consequences.

When you choose one outcome, you're doing more than narrowing your focus. You're making a statement:

"At this time and place, this is what leadership looks like."

Know where you're starting from.

Once you've named an outcome, the next step is uncomfortable but necessary—tell the truth about where you're starting.

If your goal is stabler systems, you need to acknowledge how fragile things are: how often outages happen, how close you are to failing in front of your council, board, or executives.

If your goal is a cross-department win, you need to admit the trust and relationship gaps: which departments barely speak, who doesn't believe data can actually help, or where the politics lie.

If your goal is better intake and prioritization, you need to face the current mess: emails, hallway requests, "quick questions" that turn into weeks of work, or projects no one ever formally approved.

This is the part where many leaders look away. They jump straight into action because sitting with the current state feels like failure.

It isn't failure. It's a baseline.

The leaders I interviewed who ran effective 90-day cycles treated their current reality as data, not judgment. They wrote down what was true in simple language:

- "We have two people who know how to maintain this, and both are overloaded."
- "We have five ways work can reach us, and no one is sure which one is 'official.'"
- "We have not agreed on a definition of 'urgent.'"
- "We only have 10 to 12 hours a week of real project time once we subtract meetings, firefighting, and admin."

That honesty became the foundation of their roadmap.

Choose fewer moves than you think you need.

Once you know your outcome and your starting point, the temptation is to pack as much as possible into the plan. You finally see the pattern. You want to fix everything.

Don't.

A 90-day roadmap is not a dumping ground. It is a **series of deliberate moves** that line up with your outcome.

In my own practice, I think of a 90-day roadmap like a three-legged stool: one leg stabilizes something fragile, one leg creates or advances a visible win, and the last one invests in your future capacity (time, skills, data, governance).

You might decide, for example:

- Stabilize: address the top two root causes of recurring outages or data issues.
- Visible win: deliver one cross-department map or dashboard tied to a real decision.
- Capacity: establish a basic, visible way to track work as it comes in, even if it is just a simple ticket queue.

Inside each leg, there will be tasks—people you need to talk to, technical steps, and communications. But the roadmap itself stays simple: three or four big moves, not twenty.

Several leaders told me that their breakthrough came when they stopped saying, "We'll try to work on all of this," and started saying, "We will focus on these three things, which means we will not take on X and Y right now."

That sounded risky at first. But in practice, it made it easier for executives and partners to trust them. They delivered what they said they would.

Put your roadmap in other people's language.

A 90-day roadmap that lives in your private notebook is a to-do list, but a 90-day roadmap that lives in other people's heads is a lever for change.

As soon as you have a first draft, practice explaining it without mentioning tools, apps, or internal jargon.

For example:

- "Over the next 90 days, we're going to make outages much less likely by fixing up these two weak points in our system and documenting how we respond when something goes wrong."
- "Over the next 90 days, we're going to give finance and public works a shared picture of where our highest-risk assets are, so the capital plan isn't a political fight every year."
- "Over the next 90 days, we're going to stop this chaos of requests by agreeing on one front door for GIS work and one simple way to see what's in progress."

Notice what each of those does. It names a timeframe, who benefits, and what will be different.

Once you can say it out loud, share it.

You don't have to call a big meeting. Instead, weave it into the conversations you're already having with your boss, with one or two executives, with your team, or with the colleagues who are most affected.

You're not asking permission for every detail. You're inviting alignment around the direction.

Schedule the work like it actually matters.

A roadmap is not real until it hits your calendar.

This is where the identity shift from Chapter 1 and Wendy Baldin's insight about guarding time "like gold" come back in. If your 90-day outcome truly matters, it has to appear in protected blocks of time—for you and your team, if you have one.

I've worked with leaders who built beautiful 90-day plans and then left their calendars exactly as they were. Every day was still a wall of meetings and ad-hoc requests. At the end of the quarter, they were exhausted and surprised that little had changed.

The leaders who made progress did something simple and brave—they carved out focus time first.

Sometimes, it was one morning a week when no meetings were allowed. Sometimes it was two shorter blocks while working remotely. Sometimes it meant renegotiating recurring meetings or saying, "If you want this roadmap to be implemented, something else has to move."

They also used very simple tools to keep the roadmap visible: a one-page view on the wall, a lightweight board in a ticketing system, a weekly check-in where they asked, "What did we do this week to support the roadmap?"

You don't need an elaborate project management stack to do this. You do need to treat your roadmap as something more than a best-case scenario.

Your roadmap is a promise you are making to yourself, your team, and your organization. Your calendar is evidence of whether you believe that promise.

Let the quarter teach you.

No 90-day roadmap survives contact with reality unchanged.

Emergencies happen. Key people leave. A budget cut forces you to adjust. A project you thought would be simple turns out to be more complex. Or the opposite: something moves faster than you expected.

The point of the roadmap is not to predict perfectly, but to **learn faster and more honestly**.

At the end of 90 days, instead of just rushing into the next quarter, pause and ask:

- What did we actually finish?
- What mattered most to the people we serve?
- Where did we underestimate effort or overestimate buy-in?

- What surprised us—in a good way and in a hard way?
- What does this cycle teach us about our current maturity?

These questions are the bridge into the next chapter of this book, where we'll zoom out from one quarter and talk about **strategy and maturity**.

In my own work, this is where frameworks like SlimGIM or Esri's Path to Geospatial Excellence became powerful: they gave me a way to interpret what the 90 days had taught me about where our program was fragile, where it was strong, and what sequence of change made sense next.

The tools will evolve, but the pattern holds:

- Clarify identity and vision.
- Run a 90-day experiment that fits reality.
- Let that experiment tell you the truth about your maturity.
- Use that truth to shape your longer-term strategy.

The 90-day roadmap is not the destination. It is how you walk there.

Leader's Map in Practice - Chapter 4

Pick one or two prompts. Keep it concrete; imagine you're going to act on this in the next two weeks.

1. If you had to choose **one outcome** for the next 90 days that would clearly improve how GIS serves your organization, what would it be? Write it as a single sentence that someone outside GIS could understand.

2. Describe your **starting point** for that outcome. What is fragile, messy, unclear, or under-resourced today? Where have you been underestimating that reality?

3. Sketch three "legs" of your 90-day stool: one stabilizing move, one visible win, and one capacity-building move. If you had to let go of something to make these possible, what would it be?

4. Look at your calendar. Where, specifically, will this roadmap live? What is one meeting you can renegotiate, one block of focus time you can create, or one weekly check-in you can add to keep this plan real?

In the next chapter, we'll lift our eyes again and ask a bigger question:

Now that you can run a meaningful 90-day cycle, **how do you turn those cycles into a lasting strategy**—one that is grounded in your actual maturity and aligned with the mission of your organization, not just your own energy and willpower?

CHAPTER 5:

From 90 Days to a Strategy That Lasts

THE FIRST TIME YOU finish a real 90-day cycle, something subtle shifts.

You listened. You stabilized something fragile. You shipped a visible win that mattered to people outside GIS. You protected some focus time and saw that when you did, the work moved differently.

Then the quarter ends.

You look up from tickets, meetings, and progress notes and feel a new question forming:

> *Is this it? Do I just keep doing 90-day plans forever? Or is there a bigger story I'm supposed to be telling?*

What you are really asking is:

> *Where is this all going?*

That's the work of strategy.

When vision and 90 days still aren't enough.

For a long time, my world oscillated between two modes.

In one mode, I lived in vision: big possibilities, conference inspiration, the future GIS environment I could describe in almost cinematic detail. In the other mode, I lived in survival: what can we realistically move in the next few weeks, who is yelling the loudest, what is about to fall over if I don't touch it.

Both modes were real. Both mattered. But neither, on its own, answered the bigger questions:

- Given who we are as an organization, what are we **ready** for?
- In what **sequence** should we build capabilities so GIS becomes part of the enterprise, not a side project?

- How do we avoid the pattern of trying to do everything and quietly burning out?

I knew how to describe where I wanted us to end up. I knew how to run 90-day plans. I did **not** know how to be honest about the distance between here and there.

That honesty is what maturity gives you.

The day the mirror showed up.

In 2018, I went to the Esri User Conference with the same mix of hope and exhaustion I see in many GIS leaders.

I was excited to learn, but I was also carrying the weight of a program that needed more than tuning. Systems were in place, but we didn't have a clear way to explain where we were strong, where we were fragile, and what "better" even meant in a way executives could hear.

Then I walked into a session where **Paul Giroux**, an Innovation Officer at GSU, Ontario, Canada, and Founder of SlimGIM, was doing a presentation on Slimgim, a maturity model for GIS. It broke the work down into areas like strategy, governance, data, technology, people, and engagement, and asked simple but uncomfortable questions:

- Do you have a clear, agreed GIS vision tied to business outcomes?
- Is there governance that actually works across departments, not just inside GIS?

- Do you know which data is authoritative and how it is managed?
- Are you investing in people and skills, or hoping they will keep up on their own?

As I sat there, something in me relaxed and lit up at the same time.

I wasn't crazy for feeling that "our GIS" was more than a technical environment, and I wasn't wrong for thinking we needed a structured way to talk about it.

I suddenly had a **mirror**.

When I came home and reached out to Paul, it wasn't because I wanted another framework to tape on the wall, but because I wanted a way to tell the truth.

One county GIS manager I spoke with has turned maturity assessment into a yearly ritual and a budget tool. Every year, she runs a simple capability maturity review and sits down with her leadership team to walk through the results. Instead of saying, "We're overwhelmed; we need more staff," she can point to specific capability gaps and pain points—places where the current maturity is holding the organization back. When she asks for a new position or investment, it's not a vague plea for help; it's connected to a concrete step in their growth. The assessment becomes a shared picture of reality, not a judgment.

Tools like SlimGIM gave language and shape to things I had been feeling for years. They turned vague frustrations into specific gaps and "we need to be better at GIS" into "we need to move from here to here on governance, or nothing else will stick."

If the 90-day roadmap is your first bridge between vision and reality, maturity frameworks like these are the navigation system that stops you from driving in circles.

The point is not the specific tool. It's the posture:

> *"We will anchor our strategy in evidence about where we are, not just in ambition about where we wish we were."*

Strategy as a sequence, not a wish list.

Without maturity, "strategy" easily turns into a wish list in PowerPoint.

We create serious-looking diagrams with layers and phases and lofty goals: modern data platforms, self-service analytics, real-time dashboards, digital twins. It all sounds good and even looks plausible, separated from the constraints of time, people, and politics.

Then the year starts.

Requests pour in. Outages happen. A reorg hits. The budget gets cut. The beautiful strategy deck sits on a shared drive somewhere, quietly becoming a monument to what we didn't do.

One of the things Paul has pointed out, in his writing and in our conversations, is that many strategies fail not because teams are lazy or uncommitted, but because the plan was never **real**. It was not aligned with the organization's true maturity or appetite for change. It was built to impress, not to implement.

A maturity-aware strategy looks different.

Instead of saying yes to everything, it says things like:

- "Our governance is weak, and our data is inconsistent. For the next 18 to 24 months, we will focus on establishing a few critical agreements, improving data quality in key areas, and delivering visible wins that build trust in shared data."
- "Our systems are strong, but engagement is low. For the next two years, we will prioritize co-design with departments, training, and changing how we communicate value."
- "We are relatively mature in one division and far behind in others. We will treat that division as a lighthouse—a place to pilot advanced capabilities—while we focus on basic stability and readiness elsewhere."

Strategy becomes a sequence of deliberate bets, informed by where you are now and where you're trying to go, not a vague future state you hope someone else will magically deliver.

Letting your 90-day cycles inform your strategy.

So how do you connect the 90-day world to this bigger, multi-year picture?

You stop treating each quarter as an isolated project and start treating it as data on **your maturity**.

Every 90-day cycle teaches you something:

- How much focused work can your team actually do when you protect time?
- How do stakeholders respond to change?
- Where is the technical fragility, really?
- Where is governance missing or ignored?
- Which departments lean in, and which retreat?

If, for example, you set out in a quarter to deliver one cross-department dashboard and spend most of your time arguing about whose data is correct, that is a signal: your maturity issue is not dashboards; it is data governance and shared definitions.

If you consistently fail to make progress on roadmap items because urgent requests keep blowing up your schedule, that is a signal: your maturity issue is not "team laziness"; it is intake and prioritization.

If your tech works but no one uses it without you standing over their shoulder, that is a signal: your maturity issue is not "people are change resistant"; it is engagement, training, and co-ownership.

You can ignore these signals and write another ambitious strategy, or you can let them shape your next moves.

In a maturity-aware strategy, your 90-day learnings become a feedback loop—you run a quarter based on your best understanding, you look honestly at what happened, you adjust your assessment of where you are, and you refine the next set of moves.

Over time, those cycles form a **spiral of learning** instead of a straight line of disappointment.

Turning maturity into a story, not just a chart.

Most maturity frameworks can produce charts and scores—numbers from one to five, spider diagrams, or heatmaps.

Those are useful, but they don't provide the full picture for your strategy.

The real power is in the conversations they unlock.

When we first used a maturity assessment in my own program, the most valuable moment wasn't when we saw our scores, but when we sat with leaders and said:

> *"Here's where we are strong. Here's where we are fragile. Given your priorities, here is what we propose we focus on first—and here is what we will* ***not*** *chase yet."*

We stopped talking in generic terms—"We need more GIS"—and started saying things like:

- "If we invest here first, these three risks will go down."
- "If we skip governance again, we will keep rebuilding the same things."
- "If we push advanced analytics before we stabilize these datasets, we will just create new ways to be wrong, faster."

The assessment provided evidence. The strategy translated that evidence into choices.

This is where maturity assessments resonate for me. They remind us that you can't just sprint ahead in one area and expect everything else to magically keep up. If your tech is at a level four but your engagement and governance are at level one, your strategy can't just be "more tech." It has to be, "We will pause and bring governance and people along, or the tech will collapse under its own weight."

During one of the interviews, another GIS leader described arriving in a city where three different GIS groups were all maintaining their own base maps. On top of that, a stormwater ordinance—written without any real GIS input—forced his team to find a better way to manage the data intake and distribution to other departments.

The lesson he drew wasn't "look how clever the GIS folks are," but: if we don't connect GIS strategy to policy and business rules early, we end up with technically impressive work that is strategically irrational.

In another organization, the GIS lead explained how his team could stand up web maps and dashboards in days, while a small business intelligence team needed months to deliver their work. You can imagine the tension. People started asking, "Why are they so fast and we are so slow?" It could have turned into a turf war over who "owned" dashboards. Instead, the leader in charge reframed the problem. Rather than defending GIS or criticizing BI, he presented a different proposal to senior leadership: treat spatial solutions, analytics, and integrations as a single data services function with distinct specialties.

The point wasn't to win as GIS. The point was to align everyone around the same questions the organization needed to answer, so each tool and team could play its part without spending energy protecting territory.

When you share this kind of story with executives—grounded, specific, and linked to their goals—your strategy stops sounding like a GIS wish list. It starts sounding like **risk management and opportunity design**.

Dream big, sequence humbly.

Nothing in this chapter is an argument against big dreams.

You are allowed to imagine an organization where real-time spatial insights inform every major decision, where data is trusted, where GIS is woven into the core business processes, where your team has the skills and support to keep growing.

You might even sketch that picture out: three, five, ten years from now, what does "mature" look like here?

Believe in it. Let it pull you forward.

The only thing I'm asking is that you respect the distance between here and there enough to sequence your path.

It might look like spending a year on unglamorous data work and governance before chasing advanced tools, or accepting that your first "strategy" is really a one-year hypothesis you will revisit, maybe saying no to projects that would look great on a slide but would crush your team, or being honest about what your organization can absorb, culturally and politically, without snapping back.

In the interviews, some of the most grounded leaders said versions of the same thing:

> *"We stopped trying to do everything the maturity model said at once. We picked three moves that made sense for where we were, and when those stuck, we came back for more."*

That is what a strategy that lasts feels like: not a heroic sprint, but a series of right-sized steps that accumulate into something big.

Tools will change. The work remains.

SlimGIM or Esri's Path to Geospatial Excellence—these tools are snapshots of a moment in time. New frameworks will appear, and old ones will be renamed, updated, or quietly archived.

Your strategy should not depend on any single acronym.

What needs to endure is the underlying discipline: to see clearly where you are, not just where you hope you are; to align your GIS strategy with your organization's real goals and risks; to sequence your moves based on readiness, not envy of what others have done; and to let each 90-day cycle teach you and adjust accordingly.

If you hold to that, you can swap tools in and out as the field evolves without losing your footing.

From strategy to story.

We're still in *Scale with Systems*. You've moved from identity and vision, through influence and mentors, into 90-day roadmaps and longer-term strategy.

Before we get practical with request management, governance, and data, there's one more piece we need to address:

How you talk about all of it.

Because even the most thoughtful strategy will stall if you keep describing GIS as "the map group" or "the people who say yes to everything." To move from isolated projects to a true information system, you need a different story—one that positions GIS as a system of systems, sitting alongside finance, HR, and operations.

That's where we'll go next.

Leader's Map in Practice - Chapter 5

A few prompts to ground this in your world. Pick one or two.

1. Looking back at your last 90 days, what did they **actually reveal** about your maturity in governance, data, systems, engagement, and/or capacity? What patterns do you see that you've been avoiding naming?
2. If you had to write a one-page **strategy memo** to your leadership for the next 12 to 18 months, based on where you truly are, what three priorities would you name—and what would you explicitly say you are **not** pursuing yet?
3. Think of one framework or tool you've encountered (SlimGIM, Esri's pillars, or another). Instead of trying to "do it all," what is **one** insight from that tool you could apply in the next year?
4. Imagine yourself three years from now, looking back. If your strategy had been humble about your starting point and bold about your destination, what would be true about your GIS program that is not true today? What is one small step you can take this quarter toward that version of yourself?

In the next chapter, we'll explore how to **promote the value of GIS as an information system**—not by shouting louder about maps, but by telling better stories about decisions, risk, and outcomes in the language your organization already speaks.

CHAPTER 6:

Promoting the Value of GIS as an Information System

THERE IS A MOMENT in almost every GIS leader's journey where frustration peaks.

You've done the work. You've improved the data. You've stabilized systems. You've helped departments make better decisions. And still, someone introduces you in a meeting as:

> *"This is Izabela. She does our maps."*

You smile, but inside, something sinks.

It's not about ego. It's about accuracy. You know you are running an information service that connects systems, teams, and decisions. But the story people carry about you is still "map shop."

If that gap persists, it will limit your budget, your influence, and your team's future.

This chapter is about closing that gap, not by shouting louder about GIS, but by changing the way you talk about your work—so people can finally see it for what it is: a **system of systems**, an **information system** alongside finance, HR, and operations.

Stop selling GIS. Start telling decision stories.

When people say, "They don't get the value of GIS," what they usually mean is, "They don't respond to the way I'm explaining it."

Most GIS value pitches sound like this:

> *"We have all this great data."*
> *"We can build really powerful dashboards."*
> *"We can do 3D, real-time, predictive analytics..."*

You can see eyes glaze over.

One long-time GIS leader admitted that, across his career, this was his biggest frustration: most of the organization still thought of GIS as "the maps department."

The deeper analytical work—patterns, probabilities, tradeoffs—often remained invisible. When something broke, it was obvious. When GIS quietly prevented mistakes, optimized routes, or avoided risk, it was treated like background noise. Part of his leadership journey was accepting that this frustration was real and then choosing to use it as fuel to keep telling better stories about analysis, not just prettier maps.

It's not that executives and peers don't care about data or tools; it's that they're already carrying a mental backpack full of worries: budgets, risk, regulations, staffing, service quality, politics. If you don't connect to one of those, you're just noise.

The leaders I interviewed who had broken through stopped trying to sell GIS on its own terms. They started telling **decision stories**:

- "Last winter, operations used our asset and outage maps to prioritize which lines to inspect first, reducing overtime by X percent and avoiding Y potential failures."
- "Council asked where we were investing in underserved neighborhoods. Instead of arguing from memory, we showed them a clear spatial breakdown, and the discussion shifted from blame to options."
- "We used location data to show which projects were overlapping in the same corridor, allowing us to coordinate work and cut disruption in half."

Notice what those stories do. They start with a **real-world situation**, not a feature; they name the **stakes**;

they show **what changed** because GIS existed; and they often end with a simple metric or observation.

One county GIS manager keeps a printed copy of her organization's strategic plan next to her keyboard. When she proposes a project or defends a budget line, she literally puts her finger on the goal or objective it supports. She also sends her executives short Esri case studies and articles that mirror their context—a utility story to the utility director, an emergency management example to the fire chief—with a simple note: "We could do something like this here." For leaders who don't naturally think in maps or spatial analysis, those external stories become a bridge between their world and hers.

When you repeat stories like that—in meetings, in 1:1s, in short emails—people begin to associate GIS with **better decisions**, rather than just "nice maps."

Frame GIS as an information system, not a side tool.

Language matters.

If you describe GIS as a "nice add-on" or "something we use for maps," people will mentally categorize it as an optional extra.

If you describe GIS as an **enterprise information system**, you invite a different level of seriousness.

That doesn't mean you walk into every room and announce, "We are an enterprise information system now."

It means you consistently place GIS in the same sentences as other core systems:

- "GIS sits alongside financial, HR, and asset systems as one of the ways we see and manage reality."
- "We provide a spatial lens on the same questions finance and operations are asking."
- "Our job is to integrate data across systems so leaders can see the whole picture, not just one silo at a time."

It also means you show up where other systems are being discussed. When someone talks about Enterprise Resource Planning (ERP), CRM, or asset management, you don't tune out. You ask:

- "Where does location matter in this decision?"
- "How could a shared spatial view reduce rework here?"
- "If we connected this to our GIS, what might become visible?"

In their Esri article *What It Takes to Be a GIS Leader*, **James Perdue** talks about the importance of understanding the business and speaking its language. GIS leaders who earn trust don't just advocate for their own tools. They position GIS as a way to **solve business problems**—risk, cost, service, equity—with a spatial lens.

Similarly, in another Esri article, "How to Lead with GIS," by **Alex Harper**, the argument is clear: start from the mission, not the map. Let GIS be the means, not the message.

You don't have to quote those articles in your meetings, but you can live by their principles.

Build a small bank of proof, not a giant catalog.

A common reaction when we talk about "promoting value" is to start building a huge list of everything GIS does: every app, every layer, every service.

That can be useful internally, but it's not how you change minds.

Instead, think in terms of a **small bank of proof**—a handful of well-chosen examples that demonstrate different dimensions of value: one story about saving money or time, one story about reducing risk or avoiding a bad outcome, one story about improving service to residents or customers, or a story about supporting a strategic initiative (equity, climate, growth, homelessness).

You don't need fancy graphics. You need a paragraph for each, in plain language, that you can adapt on the fly.

A GIS leader at a mid-sized county told me she keeps a one-page "value snapshot" ready all the time. When someone asks what GIS has been up to, she doesn't dump a long list, she says:

> *"Let me show you three ways GIS has helped in the last quarter."*

Then she walks them through three anecdotes tied to what *they* care about. Over time, those snapshots changed the way people talked about her team.

You can do the same.

Pick three to five "anchor stories." Refresh them as you go. Use them everywhere: in hallway conversations, leadership updates, budget memos, lunch-and-learns.

Tailor your message without losing yourself.

Promoting value is not about becoming a different person in every room, but it does mean recognizing that different audiences care about different things.

When you talk to:

- **Executives**, emphasize risk, cost, outcomes, tradeoffs, and alignment with strategic goals.
- **Department managers**, emphasize reliability, predictability, capacity, and their specific Objectives and Key Results (OKRs).
- **Field staff**, emphasize ease of use, trust in the data, and how their input changes the picture.
- **Peers in IT**, emphasize architecture, integration, security, and supportability.

The underlying truth is the same. You're just rotating it so they see the face that matters most to them.

Several leaders told me that once they stopped using the same GIS pitch everywhere and started tailoring it, the resistance eased. People felt seen. They stopped hearing, "Here comes GIS to tell us what we should care about," and started hearing, "Here is someone trying to help me with my real problems."

You don't have to script this perfectly. You can start simple:

- "For you, what does success look like?"
- "The part of GIS that matters most to your work is..."
- "Here's one way we could make your next decision easier..."

Everything else is detail.

Make wins visible without bragging.

For some of you, all of this may sound dangerously close to self-promotion. You might worry about being seen as arrogant or pushy.

There is a difference between bragging and making value visible.

Bragging centers you: *your* brilliance, *your* heroics, *your* stress. Making value visible centers the **work** and the **impact**.

One way to check yourself is to listen to your pronouns:

- "I built..." leads one way.
- "We were able to help..." or "This allowed the team to..." leads another.

You can also invite others to tell the story. Ask a department head:

> *"Would you be willing to share at the next leadership meeting how that new map changed your planning conversation?"*

Let them be the hero. You can be the architect in the background. Your influence grows either way.

This is also where artifacts help: a simple before/after map, a screenshot, a short testimonial, a one-page case study. They do not have to be perfect. They just need to exist and circulate.

Over time, those visible wins create a new default story:

"Oh, GIS—they're the ones who helped us see X and decide Y."

That's the story you want.

From value stories to executive support.

We are still in the *Scale with Systems* phase. You've built a vision, a 90-day roadmap, and the beginning of a longer-term strategy. Now you're learning to talk about GIS as an information system, using decision stories and tailored language.

The next step is to bring that same clarity into the rooms where resource decisions are made.

It's one thing to talk about value with peers; it's another to sit across from an executive and ask for time, budget, or political cover.

To do that well, you need more than enthusiasm. You need a simple way to show them where you're headed, how you'll get there, and what they can expect if they provide the support you seek.

That's where we'll go next: building **executive support** that lasts.

Leader's Map in Practice - Chapter 6

Pick one or two. Treat them as drafts you can refine over time.

1. Write one **decision story**: a real situation, the stakes, what GIS did, and what changed. Then rewrite it in language a non-GIS colleague would use.
2. In one paragraph, describe GIS as an **information system** in your organization, sitting alongside other core systems. No jargon. Who will you test that description with this month?
3. Choose three "anchor stories" that show different kinds of value (cost/time, risk, service, strategy). How could you share one of them in the next leadership email, staff meeting, or even a hallway conversation?

CHAPTER 7:

Gaining Executive Support

TALKING ABOUT THE VALUE of GIS in the hallway is one thing. Sitting in front of an executive and asking for real support is another.

For many GIS leaders, those moments trigger an old story:

> *"I'm just the technical person."*
> *"I don't speak their language."*
> *"If I ask for too much, they'll shut me down."*

So we play it safe. We present too much detail. We bury our ask. We hope that if we show enough impressive maps, executives will connect the dots on their own.

Most of the time, they don't. Not because they're uninterested or hostile, but because their world is full of competing priorities, risks, and noise. If you don't help them see exactly where GIS fits, your work becomes background.

This chapter is about changing that dynamic.

Executives are human (overloaded humans).

The first shift is to remember that executives are human beings with overloaded calendars and very specific worries.

They think in terms of risk they are responsible for, money they have to justify, political and public pressure, and commitments they've already made.

They are not awake at night wondering, "How can I help the GIS team feel more fulfilled?" They are wondering:

- "Are we going to blow this budget?"
- "Will this failure land on my desk?"
- "Can we show progress on the strategic plan?"
- "What is going to surprise me next quarter?"

Your job is not to dazzle them with GIS. Your job is to **make their thinking easier**.

That begins with curiosity. Before you present anything, you ask questions like:

- "When you think about the next year, what worries you most?"
- "What would a 'good year' look like from your seat?"
- "Where do you feel like we're flying blind?"

Then listen. Don't tell them how GIS solutions might save them. Just listen.

Only after you understand their mental map do you start talking about how GIS can help.

The three decisive slides.

In my work with leaders, I often talk about the power of **three decisive slides** when engaging executives:

1. **The North Star slide**—what GIS is for.
2. **The 90-day roadmap slide**—what we're doing next.
3. **The value snapshot slide**—what difference it's already making.

You don't have to use literal slides. Think of them as three essential messages.

The North Star slide answers:

> *"In this organization, GIS exists to improve these kinds of decisions in these parts of the business."*

It might show a simple diagram: organizational priorities on one side, GIS capabilities on the other, and where they intersect. No acronyms. No architecture diagrams. Just a clear link between mission and GIS.

The 90-day roadmap slide answers:

> *"Here is what we are focused on in the next quarter, and why."*

It shows three or four initiatives, each tied to a real outcome: a more reliable system, a specific cross-department win, a new governance agreement. It makes tradeoffs visible: "We are not chasing X yet so that we can deliver Y."

The value snapshot slide answers:

> *"Here is what has changed so far because of this work."*

It offers a handful of before/after statements, like the ones you developed in Chapter 6: reduced time to answer common questions, fewer outages, clearer decisions, or actions taken because of GIS insights.

Together, these three pieces do something important:

They show executives that you have **direction** and discipline and that you produce **results**.

You are not asking them to bet on a vague dream, but inviting them to invest in a structured, evidence-based approach.

Ask for support like a partner, not a supplicant.

Once you can tell that story, you can be more direct about what you need.

Instead of hinting or hoping, you can say things like:

- "To deliver this 90-day roadmap, we need X hours of protected time and a Y level of responsiveness from these departments."
- "To reduce this risk, we need to invest in stabilizing this part of the system. That will cost approximately ___ and will avoid ___."
- "To make governance real, we need you to endorse this committee and show up to the first meeting."

Notice that you are not just asking for money, but for **decisions, attention, and air cover**.

In the *How to Lead with GIS* article, one of the underlying messages is that successful leaders don't wait at the edge of the table hoping to be noticed. They step forward with a point of view and a specific ask, grounded in the organization's goals.

You can do this without arrogance, by saying:

> *"From my seat, here's what I see. Here are two options. Here's the risk of doing nothing. What feels most aligned with your priorities?"*

You are positioning yourself as a partner, not a petitioner.

One GIS manager said early on: "If they don't see us, it's easier to forget us." She shows up everywhere—GIS Day events, Board of County Commissioners meetings, public proclamations, and community outreach. Sometimes she presents; sometimes she's just in the room, listening and learning. Over time, her presence has shifted how leaders talk about GIS. It is no longer an invisible line item; it's a face, a voice, and a set of capabilities they associate with real community outcomes.

If you are an introvert, none of this means becoming a different person.

I say that as someone who still needs to retreat to a quiet place to restore my energy. For most of my career, I would have chosen a day of focused map work over a day of meetings with executives. But to grow the program and make it visible, I had to learn how to get out of my shell on purpose. Presenting was one of the things that helped me most. When I treated every presentation as a chance to tell one clear story about value—not to perform, but to serve—it became easier to stand up in front of people who scared me a little.

Several of the leaders I interviewed described a similar pattern. One county GIS manager told me that they are naturally introverted but "put on a hat" when it is time to speak to commissioners or senior leaders. They prepare carefully, use strong visuals, and often borrow short Esri case studies or examples from other jurisdictions, so they are not starting from a blank page.

Another GIS professional in a small county auditor's office blocks time afterward to be alone, to think and recharge, because they know those external-facing moments cost energy. The common thread was not pretending to be extroverted, but being willing to step forward when it mattered—and then giving themselves the recovery they needed so they could keep doing it.

If this is you, start small. Ask for ten minutes on an agenda instead of an hour. Share one clear example instead of a full tour of everything your team does. Script your opening sentence and closing sentence so you know how you will begin and how you will end. Use your natural strengths: introverts are often excellent listeners and pattern-finders. Bring those gifts into the room. You do not have to become the loudest voice at the table to lead. You do have to be willing, at least sometimes, to let people see you.

Budget as a story about risk and value.

Budget conversations can be intimidating, and many GIS leaders feel like they're constantly defending line items:

- "Why do we need this license?"
- "Can't we delay this upgrade?"
- "Why do you need another person?"

If you stay at the line-item level, you spend your energy arguing about prices instead of value.

A more strategic move is to frame the budget in terms of **risk and outcomes**.

For example:

- Instead of, “We need $X for this software,” you might say, “Right now, our outage risk is at this level because of these factors. Investing $X here reduces that risk and protects this revenue/service.”
- Instead of, “We need another analyst,” you might say, “We are currently handling this volume of demand with this capacity, resulting in delays for these priority projects. Adding one role would allow us to deliver these three outcomes this year.”

None of this guarantees a yes. Executives still have to juggle competing demands, but it reframes the conversation in their language: risk tradeoffs, opportunity costs, and visible outcomes.

It also builds trust: you are clearly thinking about the organization as a whole, not just your own wish list.

Trust is built in the small things.

Executive support isn’t won in a single big presentation. It’s built—or eroded—in the small interactions before and after.

Several leaders I spoke with said their relationships with executives changed when they started doing three simple things consistently:

- **No surprises.** When something went wrong, they told the executive early, with a plan, instead of hiding it and hoping it would blow over.
- **Closed loops.** When an executive asked a question or requested something, they followed up—even if the answer was "We couldn't do that yet, and here's why."
- **Visible wins.** When something went well, they sent a short note: "Remember that thing you supported? Here's what it just made possible."

Over time, executives stopped seeing them as "the GIS person who shows cool stuff sometimes" and started seeing them as **reliable partners**.

In *What It Takes to Be a GIS Leader*, Perdue emphasizes relationships and trust as the core of GIS leadership. The same message came through in the interviews: what executives remember isn't every map you showed them, but whether you kept your word and helped them sleep a little better at night.

Don't put all your weight on one champion.

One caution: it is tempting to latch onto a single supportive executive and build everything around that relationship.

Sometimes you have to start there. One sponsor is better than none.

But organizations change. People move, get promoted, retire, or burn out. If all your support is tied to one person, your work becomes fragile.

As your influence grows, think in terms of a **small coalition**: one executive sponsor, one or two peer champions in key departments, one person in finance or budgeting who understands your story, someone in IT who sees GIS as part of the core.

You don't have to formalize this as a "coalition" on paper. You build it through conversations, shared projects, and consistent behavior.

Then, when someone leaves, you are not starting from zero.

From the executive suite back to the inbox.

At this point in the book, you have reframed your identity, clarified your vision, mapped your influence and support, built a 90-day roadmap and a longer-term strategy, started telling better value stories, and begun engaging executives as partners.

The next reality check is this—none of that will stick if, on Monday morning, work still arrives in a chaotic pile and no one can see what you're actually doing.

To keep the promises you make—to yourself, your team, and your sponsors—you have to redesign how work and data flow through your world.

That is where we're going next: from **inbox to impact**.

Leader's Map in Practice - Chapter 7

One or two of these is enough; pick what feels most alive for you right now.

1. Imagine you have 15 minutes with a key executive. Draft the three messages you would want them to walk away with: your North Star, your next 90 days, and one value snapshot. No slides, just language.
2. Think of one upcoming or recent budget conversation. How could you reframe your request in terms of risk and outcomes instead of line items? Write that version down.
3. Who are your current or potential champions at the executive and director levels? Name at least three people. What is one small action you can take in the next month to strengthen those relationships?

CHAPTER 8:

Inbox to Impact—Designing Your Front Door for Work

YOU CAN HAVE A beautiful strategy, executive support, and a clear vision...and still feel like you are drowning by Wednesday.

Everywhere you turn, something is buzzing—an email with the subject line "Quick Map?" that turns into two weeks of work; a Teams message during another meeting: "Do you have a second?"; a manager catching you in the hallway: "Could you just pull some data for me?"; a director dropping by with a "high-priority" request no one else knew about.

By Friday, your calendar is a graveyard of good intentions. You were busy every hour, yet when you try to connect the week back to your 90-day roadmap, it's hard to see what actually moved.

This isn't a management issue. It's a leadership issue.

If we don't design how work enters our world, someone else will design it for us.

Chaos is not responsiveness.

For a long time, I equated responsiveness with saying yes to everything.

If someone reached out, I wanted to help. I cared about relationships, and I wanted GIS to be known as the team that "gets it done." Especially after being the lone GIS person, it felt almost like a moral obligation.

The unintended result was predictable: our days were full of interruptions, important work kept slipping, the loudest voices got the most attention, and no one—including me—had a clear picture of our workload.

We weren't leading. We were reacting politely.

It took me a while to admit the truth:

> *Letting work come in from every direction, at any time, in any format, is not kindness—it's abdication.*

Once you begin to see yourself as an information leader and a steward of your team's energy, you cannot keep pretending that chaos is "just how it is." You have to decide what kind of system people are stepping into when they reach out to GIS.

That is what a front door is for.

One front door, many conversations.

A front door is not a piece of software. It's a promise:

> *"Here is how you access our help. We will treat your requests seriously and fairly. In return, we will not let everything pile onto individuals in invisible ways."*

In practice, this usually means choosing **one primary path** for requests: a ticketing system, a simple request form, or a shared intake queue.

When I led a GIS program, we used a basic ticketing tool and a short morning check-in. Every weekday, we'd stand together (or meet online) for a few minutes:

What I did yesterday.

What I'm working on today.

What I need help with.

If someone mentioned a new request in the hallway, I'd say, "Great, can you put that into the queue so we don't lose it?" Not because I loved the process for its own sake, but because I knew what happened when we relied on memory: things slipped, people burned out, and trust eroded.

If you are a team of one, the same principle applies. The front door might be a simple form or a spreadsheet only you see. The point is that every request lands somewhere, so you can see your true load and make conscious choices instead of being pulled by whoever asks last.

This isn't about building a wall. It's about putting a handle on the door.

Making invisible work visible.

One of the biggest shifts that happens when you create a front door is that invisible work becomes visible.

You begin to see how much time goes to "quick" requests, how many things never had a clear owner or deadline, and, finally, how many projects started without anyone agreeing on what success meant.

This can be sobering, but it can also be incredibly freeing.

Instead of telling your director, "We're really busy," you can say:

"Here's what's currently on our plate, here's what came in this week, and here's how that affects the roadmap you asked for."

Now you're not complaining. You're providing information they can act on.

If you're leading a team, it has another effect: people stop quietly suffering under private piles of work. They see that everyone is carrying a share, and they know it's okay to say, "I am at capacity."

You go from a culture of hidden heroics to a culture of visible stewardship.

Prioritization as an act of care.

Once you have a clearer view of requests, you inevitably run into a hard truth:

> *There is more demand than capacity.*

That's not a sign that you're failing. It's the reality of almost every modern organization.

What matters is **how** you respond.

During one of the interviews, one leader shared with me the time when he noticed a different kind of risk: when you have smart GIS people with broad access to servers and tools, it's very tempting to start solving everything on the spot. A department describes a problem in a meeting, and part of you thinks, "We could build something for that by the end of the week." That speed feels great in the moment, and stakeholders love it, but over time it quietly turns GIS into a rogue mini–IT shop, operating outside shared priorities and capacity limits.

He made a deliberate choice to route GIS work through the same intake and prioritization process as other technology requests, even when their team could have moved faster alone. It was an act of stewardship: protecting the long-term health of the environment and the trust of their partners, rather than chasing short-term applause for being the team that always says yes and delivers overnight.

Before I had any structure, I tried to keep everyone happy. I overcommitted. I said yes when I should have said "not now." I let urgent-but-low-impact work shove important work aside.

From the outside, it looked like dedication. On the inside, it was unsustainable.

A healthier approach is to treat prioritization as an act of care for your team, your stakeholders, and your commitments.

You begin to ask—does this request support our current 90-day outcome or strategic priorities? Is this truly an emergency, or just urgent for one person? Can we solve part of this with a smaller effort now and revisit the rest later? If we accept this, what will have to move?

Sometimes that leads to a hard but honest response:

> *"We want to help with this, and it is important. If we add it this quarter, these two things will slip. Which is more critical from your perspective?"*

Or:

> *"We can't build the full solution in the next month, but we can meet with you to clarify the problem and sketch options."*

Or, very simply:

> *"We don't have capacity for that right now. Let's put it on the list for when we plan the next 90 days."*

You will not always get the answer you want, but you will be leading, not apologizing.

Respecting capacity as a real constraint.

Many of the leaders I interviewed, and many who responded to the "From Technical to Leadership Journey" survey, described the same pattern: people expecting them to operate as if they had unlimited time.

Only once they got honest about capacity did things begin to change.

For some, that meant sitting down with their team and calculating:

> *Once you subtract meetings, support, emergencies, and admin, how many hours per week are left for project work? How many real, focused chunks of time do you have?*

The number is almost always smaller than you think.

When you know that number, your front door and your prioritization suddenly feel less theoretical. If you have, say, 40 to 50 hours a week of project time across the whole team, you start treating that time as something precious, not as a vague resource to be stretched.

You become more realistic in your 90-day roadmap. You stop quietly blaming your team for "not moving fast enough" when the real issue is that you've promised three times more work than anyone can reasonably do.

This isn't lowering the bar. It's designing for reality so you can actually deliver.

Teams of one, many voices.

If you are a team of one, you might be thinking, "This all sounds great for bigger teams. But what about me?"

The principles still apply—in many ways, they're even more important.

A simple request log or board becomes your backup memory. A weekly check-in with yourself or your manager becomes the place to renegotiate priorities. Saying "not now" becomes a way to protect your only asset: yourself.

Several people I interviewed started their leadership journey exactly there—as the only GIS person in a department. They didn't wait for a larger team before designing some kind of system. They used whatever tools they had to track work, show their boss what was on their plate, and gently push back when expectations didn't match capacity.

Those early habits made it easier to build a healthy team later, rather than replicating the hero pattern at a larger scale.

From front door to rules of the game.

Designing a front door and basic request flow is the operational side of Scale with Systems. It gives shape to your days and weeks. It protects your roadmap from being swallowed by noise.

But there is a deeper layer we haven't addressed yet:

Who decides what gets on the roadmap in the first place?

Who owns the data behind your maps?

How do you avoid the "whose numbers are right" fight in front of the council or the board?

How GIS decisions tie into broader IT and organizational governance.

In other words, what are the **rules of the game**, and who writes them?

That is the work of governance.

In the next chapter, we'll move from the flow of requests to the agreements and structures that make that flow trustworthy—for projects, time, and data.

Leader's Map in Practice - Chapter 8

Pick one or two of these and answer in complete sentences, not bullet points.

1. How many different ways can work currently reach you (email, chat, hallway conversations, forms, systems)? What would it look like to move toward one primary **front door** over the next 60–90 days?
2. Think about your true capacity for project work in a typical week. Once you subtract meetings, support tasks, and emergencies, how much focused time is left? How does that change what you can honestly promise in your next 90-day plan?
3. Write one sentence you could use to say **"not now"** that still respects the relationship. For example: "I want to give this the attention it deserves, but right now we're at capacity until ___. Can we revisit it when we plan the next 90 days?"

CHAPTER 9:

Governance That People Actually Use—Projects, Time, and Data

THE WORD "GOVERNANCE" DOESN'T usually make people smile.

It can sound heavy and bureaucratic, like endless meetings and documents no one reads. Many GIS leaders have seen governance attempts fail: committees that meet twice and disappear, standards that live in a shared folder while everyone keeps doing what they were doing before.

And yet, when you listen to the stories behind the frustration, governance is exactly what's missing.

Projects stall because no one is sure who decides. Priorities shift with every new email from above. Teams argue in front of leadership about which data is right, and GIS is blamed for the inconsistency it never had the power to fix.

Governance is not a punishment for success. It's what you build if you want your success to last.

Governance, in plain language.

Underneath the jargon, governance is just this:

> *"The way we agree to make decisions about shared things."*

In GIS, those shared things include projects that affect multiple departments, data that many people rely on, standards that keep systems from breaking, and the limited time and capacity of your team.

Without governance, every decision becomes a one-off negotiation, where whoever is loudest or closest to power tends to win, and people who depend on stability end up exhausted.

With even a little governance, you move some of those decisions out of private channels and into the open; out of personalities and into agreed principles; and out of wishful thinking and into realistic sequences.

In a small county with no formal GIS department, one practitioner described how she built data governance out of simple agreements. There was no grand council, no thick charter. Instead, she drafted memoranda of understanding that spelled out who owned which datasets, who was responsible for updates, and how changes would be communicated. It wasn't fancy, but it was clear. When questions arose about whose data was right, they had something concrete to point to. Governance, in that context, was not a big meeting, but a set of promises on paper.

In my own leadership journey, life got easier whenever we paused long enough to ask, "Who should decide this? Who is affected? How do we want to handle similar situations going forward?"

It didn't happen magically overnight, but it helped us breathe easier.

A lightweight house, not a palace.

You do not need a giant committee structure or a 40-page policy to have real governance. In fact, too much structure can be just as bad as too little; people smile, sign, and then go back to improvising.

A better image is a small, sturdy house: just enough walls to know which room you're in, clearly marked doors, a roof that protects from the worst weather, and room to add on later if needed.

For many organizations, that looks like a **cross-functional GIS or geospatial steering group**. It might meet monthly or quarterly and include someone who speaks for operations or public works, someone from planning and development, someone from IT, someone from finance or budgeting, and you, representing the GIS function.

The point is not to micromanage every ticket, but to agree on priorities for bigger initiatives, resolve conflicts between departments, endorse and communicate standards, and connect GIS decisions to the broader business and IT governance.

In one interview, a leader described the before-and-after this way:

> *"Before, everything was back-channel deals. Projects started because someone knew someone. After we set up a simple steering group, people might still lobby, but the actual decisions happen in the open. It took the pressure off me as the middle person."*

You can start small. Even three people meeting consistently to talk about shared GIS impacts is better than carrying everything alone.

When governance is missing, your team pays.

Without governance, your 90-day roadmap and your front door are constantly under attack.

A director insists that their project is a "top priority" and must jump the queue. Another department quietly hires a vendor to build a parallel solution. Someone changes a critical dataset without telling anyone. A "quick favor" for a senior leader diverts half your capacity.

You are put in the role of referee and rescuer.

With governance, you have somewhere to take these tensions. You can say:

> *"This is exactly the kind of conflict we said we'd handle together. Let's bring it to the steering group."*

You are no longer the sole gatekeeper. You're part of a group making tradeoffs on behalf of the organization.

For your team, this matters. They see that decisions are not random. They see that when they say, "We're at capacity," you have a venue to back them up. Governance becomes one way you **protect your people**, not a way to control them.

Data governance: one version of the truth.

If project governance is about what you work on and in what order, **data governance** is about what everyone is allowed to call "true."

Many GIS professionals have experienced a version of this moment:

You are in a high-stakes meeting. A map appears on the screen. Someone asks a simple question: "How many of X do we have in this area?" You answer based on the layer you maintain.

Then someone else pulls up a spreadsheet or a different map and says, "That's not what I have." The room tilts. Suddenly, the conversation is no longer about the decision; it's about whose data is right.

The issue isn't just technical. It's about trust.

Data governance is the set of **agreements and habits** that changes this pattern. It answers questions like which datasets are considered authoritative for specific purposes, who owns and maintains each of those datasets, how changes are made and communicated, what minimum level of documentation or metadata is expected, and how historical versions are handled, so the changes can be explained over time.

You don't need a full-blown enterprise data program to make progress. You can begin with a handful of critical layers: assets that drive big spending, boundaries that shape policy, or customer or resident data that affects services.

For each, you convene the people who care most and ask:

- "Who is willing to own this?"
- "What decisions does it support?"
- "What does 'good enough' look like for this layer?"
- "How will others know what they can rely on?"

In the maturity work developed by **Paul Giroux** and others, data governance is a core pillar. Without it, advanced analytics and flashy apps simply become new ways to be wrong—faster.

In everyday terms, if you consistently have the "whose numbers are right" argument, your next big initiative shouldn't be a new dashboard, but rather data governance.

Governance that fits your culture.

Every organization has its own personality.

Some are comfortable with formal charters and committees, while others rely heavily on relationships and informal agreements. Good governance respects that culture instead of trying to import a foreign one.

If your culture is informal, your first governance step might be a regular breakfast with a few key departments where you agree on priorities and ownership, or a short, plain-language document that says, "For these top five layers, here's who owns what."

If your culture is more formal, you might anchor your GIS governance into existing structures (IT governance, project portfolio boards) or propose a simple charter for a geospatial steering group that plugs into them.

One leader I spoke with realized that even solid data governance wasn't enough. The organization was getting better at structured data—authoritative layers, clear owners, known update cycles—but decisions were still being made with information buried in documents, slide decks, emails, and shared drives. So they expanded governance to cover knowledge, not just data, and created simple rules for how reports were stored, how key decisions were documented, and how people could find the "last, best version" of something without digging through folders.

It wasn't a fancy knowledge management program, but a recognition that if governance stops at tables and feature classes, the story is incomplete. People make choices with whatever information they can reach. Good governance makes the right information easier to reach, wherever it lives.

In both cases, the test of whether governance is working is not what's on paper, but whether fewer things blindside you, fewer arguments erupt about data in front of leadership, or your roadmap and capacity are respected more consistently.

If the answer is yes, you are on the right track. If not, adjust. Governance is not a one-time design. It's something you tune as you learn.

Rules that serve people.

It is easy to turn governance into an excuse:

> *"We can't do that, the policy says no." "We have a committee for that. Good luck getting on the agenda."*

That is not the kind of leadership this book is about.

Governance in the spirit of **Serve with Stewardship** has a different flavor. It asks:

- Are our rules actually helping people make better decisions?
- Are we reducing confusion, or creating more?
- Are we protecting what matters, or just protecting ourselves?

The leaders in your organization—executives, managers, frontline staff—do not need more process for its own sake. They need clarity, fairness, and reliability. They want to know:

- "If I invest time with GIS, will it go somewhere?"
- "If I trust this map, will I be embarrassed later?"
- "If we agree on a standard, will it stick?"

Governance is one of the ways you answer yes.

From structures to people.

With this chapter, you have the core of **Scale with Systems**: 90-day roadmaps and realistic strategy, value stories and executive support, a front door for work that protects capacity, and governance for projects, time, and data that people can actually use.

All of this is necessary. None of it is sufficient.

Systems are always run by people: overworked analysts, newly promoted managers, skeptical directors, and field staff who have seen many initiatives come and go.

In the next part of the book, we turn toward how you hire the right people for where you're going; how you build a team, not a hero factory; and how you align and grow your people so they can thrive inside the systems you're building.

That is the heart of **Serve with Stewardship**.

Leader's Map in Practice - Chapter 9

Choose one or two prompts and answer them honestly. Think names and situations, not abstract ideals.

1. Think of one recurring conflict or confusion you've seen around GIS projects or data in the last year. If you treated it as a **governance problem** rather than a personality problem, what agreement or structure might have prevented it?
2. List three datasets that are critical for big decisions in your organization. For each, who is (or should be) the owner? What is one small step you can take in the next 60 days to strengthen data governance for just one of them?
3. If you were to invite 4–6 people into a first, light-weight GIS/geospatial steering conversation, who would they be, and what's the **first decision** or point of tension you'd want that group to address together?

Part III:
Serve with Stewardship

EVEN THE BEST SYSTEMS will eventually crack if the people inside them are exhausted, confused, or afraid.

Serve with Stewardship is about the human side of GIS leadership: how you hire, grow, and protect the people who do the work—including when the "team" is just you and one other person.

Stewardship is different from heroics. A hero rushes in to save the day, while a steward asks, "What conditions do my people need so they can consistently do their best work?" and then deliberately builds those conditions.

In this part, we will explore:

- Hiring the right people for where you are now and where you are going, not just the person who looks best on paper.
- Building a team, not a hero factory—moving away from a culture where one or two stars carry everything while everyone else quietly burns out.
- Creating role clarity and growth paths, so your analysts and specialists can see a future for themselves without leaving GIS or your organization entirely.
- Earning buy-in and alignment, so your team understands and believes in the vision, not just the next ticket in the queue.
- Helping your team define its own North Star, so they are motivated by more than just "keeping the lights on."

You will hear from leaders who were explicitly told, "Make these people a team," and from those who had to build a sense of team across silos, contractors, and remote staff. You will see how they used mentoring, stretch assignments, and honest conversations to turn groups of individuals into something more resilient and generous.

Stewardship does not mean being "nice" at the expense of standards. It means holding both care for the people in front of you and a commitment to the quality and integrity of the work.

In the end, your maps, apps, and dashboards are only as strong as the relationships, trust, and shared purpose underneath them. This part helps you build those foundations so that when pressure comes—and it will—your team bends, adapts, and grows instead of quietly breaking.

And there is one more layer. All of this—identity, systems, team—rests on the person doing the leading, which brings us to the inner work.

CHAPTER 10:

Hiring the Right People

AT SOME POINT IN your leadership journey, you realize a hard truth: you can't outwork a bad system, nor can you outwork a bad fit on your team.

You've reframed your identity, clarified your vision, built 90-day roadmaps, started shaping strategy, and designed a front door and governance to keep the chaos in check.

Now the question becomes: who is going to help you run all of this, and what kind of people do you want to build it with?

This chapter is not about hiring in the HR sense—posting a job, checking boxes, filling a seat.

It's about **designing a team that can carry the vision without burning out** and about being honest enough to hire for what you actually need, not just what's comfortable.

Leadership is not a solo sport (even when you're still a team of one).

Many GIS leaders start their journey as a **team of one**.

One person who knows the software. One person whom people email when they need a map. One person who stays late to fix broken services or update data because no one else can.

During that time, it's easy to internalize a story:

"If I don't do it, it won't be done right."

That story may even have been true at the beginning, but if you carry it too long, it becomes a trap.

It makes you the bottleneck, keeps others from growing, and quietly tells your brain, "You are safer staying the hero than building a team."

Several leaders I spoke with described the moment they realized that staying the hero was actually **breaking** the program. Work stalled when they got sick or went on vacation. Projects depended entirely on their personal knowledge, and there was no slack for thinking, only reacting.

Hiring—or reshaping—a team is one way you move from **hero** to **builder**.

Even if you're still a team of one, this chapter is for you because the way you think about your future team shapes how you make decisions today: what you automate, what you document, and what you're willing to let go of.

You are hiring for a journey, not a task list.

Most job descriptions are written backwards.

They start as a long list of tasks and technologies: must know X, Y, Z; will maintain A, B, C; will support these systems; will respond to those tickets. By the time you are done, you've described a person who doesn't exist.

Or worse, you've described another version of yourself.

Brandi Rank, a Principal Product Engineer at Esri, challenged me to look not just at the description but at the title itself. As long as our business cards say "GIS Analyst" or "GIS Technician," people will keep seeing us as the map people in the corner. The actual work might be business analysis, decision support, or service design, but if we wrap it in a narrow technical title, we quietly box ourselves in. Her point wasn't that titles magically change everything, but that they signal how we see the role and how others are invited to see it. When you are hiring, you are not just choosing a person; you are choosing a label for the kind of work you expect them to be part of. Sometimes the most powerful move you can make is to name the role for its impact—business analyst, product owner, information services lead—and let the title open a bigger door than "the GIS person" ever could.

When you think like a leader, you approach hiring in a different way. You start with questions like: where is this program in its journey; seeing GIS as an enterprise information system, what are we trying to build in the next 2 to 3 years; what am I uniquely good at, and what skills do I need someone else to bring; or where is our program mature, and where does it need new energy?

If your program is still in early stages—stabilizing systems, cleaning foundational data, proving value—you may need someone who is comfortable rolling up their sleeves, doing hands-on work, and dealing with ambiguity.

If you are further along—with stronger infrastructure, clearer governance, and rising demand—you may need someone who loves integration, coordination, communication, training, or analytics more than deep configuration.

You are not hiring a generic "GIS person." You are inviting a **specific kind of partner** into a specific chapter of the story.

When you write or refine a role, try adding a sentence like:

"In the next 18 to 24 months, this role will help us move from ___ to ___."

That one sentence can change how you and your candidates see the work.

Traits over tools.

Tools matter. You need people who can actually do the work.

But in almost every interview and survey response, when leaders talked about their best hires, they talked less about specific certifications and more about traits.

They used words like curious, dependable, willing to learn, and kind under pressure; they valued honesty when someone was stuck, an ability to talk with non-technical people, and a lack of resentment toward or feeling threatened by someone else's success.

They talked about people who cared as much about **the mission** as the map. People who would ask, "What problem are we really solving here?" not just, "What symbology do you want?"

Many also talked about the cost of hiring the wrong traits: technically brilliant but dismissive of others; endlessly clever but allergic to documentation; willing to say yes in the interview but "that's not my job" once hired.

One leader told me that the turning point in their hiring came when they started asking themselves three questions:

- "Would I trust this person to talk directly to my most skeptical director?"

- "Would I be comfortable sitting in a car with them for three hours on the way to a site visit?"
- "Do they light up when they talk about helping others, or only when they talk about their own projects?"

You don't have to use those questions. But you do need some way to check for alignment on **how** someone works, not just what they know.

Skills can be taught. Character and curiosity are much harder to retrofit.

Don't hire clones of yourself.

There is a real temptation, especially with your early hires, to look for smaller versions of yourself.

Same background. Same strengths. Same way of solving problems.

It feels safe. You can speak in shorthand. You can "just know" they'll do it your way.

I know because I've done it.

But over time, that safety became a weakness.

I ended up with a team that thinks the same way and enjoys the same kind of work. Whole categories of work got neglected: facilitation, documentation, testing, relationship-building, and training. In the end, I created a culture in which people felt they had to lead like me, instead of bringing their own strengths.

Once I realized the shortcomings of my own doing, I decided to make a shift and try something new. With time, I developed a system that later on became part of my leadership training, where I talk about roles in broader terms:

- builders and maintainers,
- translators and facilitators,
- explorers and stabilizers.

You don't need one of each with official labels. But you do need a mix.

Maybe you're naturally a visionary builder who loves starting new things. Then you might look for someone who enjoys refining, finishing, and keeping systems healthy.

Maybe you are strong at pattern recognition and system design but less comfortable in high-stakes conversations. Then you might look for someone who can be a calm voice in front of departments or leadership.

The point is not to replace yourself. The point is to **design a team** that can handle the work your vision creates.

Hiring when you're still a team of one.

For many readers, the first big hiring decision is very specific:

> *"I'm finally allowed to hire one more person. What do I do?"*

You feel the weight of that choice. Get it right, and your life and impact will change. Get it wrong, and you'll spend a year exhausted and frustrated.

Here are a few questions to consider before you write the posting:

- What am I doing today that **only I can do?**
- What am I doing today that someone else could absolutely do, given the right support?
- If I keep doing everything I'm doing now, what will break first?
- What will we need more of in 12 to 18 months from now that we currently don't have?

Your first hire is rarely "another you," even if that's what you're shooting for. Often, they are someone who can take on operational work so you can lift your head; someone who can specialize where you are a generalist; or someone who is hungry to grow into spaces you're ready to share.

It also means being honest with them in the interview:

> *"We are still building this plane while flying it. Some days you'll be configuring systems. Some days you'll be helping me think through strategy. Some days you'll be troubleshooting things that weren't really your fault. If that excites you more than it scares you, we're going to get along well."*

Clarity like that doesn't scare away the right person. It attracts them.

Sometimes you start by hiring temporary help—an intern who can help you get your feet under you.

One leader fought for years to get funding for GIS interns. When he finally did, he treated the role as more than cheap labor. Interns took on real projects, presented their work, and saw how it connected to the city's mission. His last intern eventually took his job when he retired. She didn't arrive as a fully formed expert; she arrived as someone curious, capable, and willing to grow. Because he had built a culture that valued learning and shared ownership, promoting her into his role felt natural.

Hiring for potential and then actually giving people room to grow is one of the most strategic things you can do as a leader, especially if you have a limited budget.

When you inherit a team you didn't hire.

Sometimes you don't get to build a team from scratch. You inherit one.

You walk into a role, and there are already people with titles, histories, habits, and stories. Some may have been waiting for a leader like you. Some may resent that you took their opportunity to lead. Most will be watching carefully to see what kind of leader you are.

In that situation, your first job is not to judge. It's to see. It's to listen. You ask:

- "What are you proudest of in the last year?"
- "Where do you feel stuck?"

- "What kind of work do you wish you could do more of?"
- "What do you wish leadership understood about your role?"

You watch how people respond to stress—who steps up, who disappears, who sits quietly, patiently waiting for this "inquisition" to be over, and who needs more support, not more pressure.

Then, over time, you start to shape roles around strengths and the needs of the program.

That might mean shifting responsibilities so one person is more focused on automating workflows and another on analytics. It might mean slowly moving someone out of a role that no longer fits and helping them land somewhere better. It might mean coaching someone who has the right heart but needs new skills.

You will be tempted to think, "If only I could wipe the slate clean." That temptation is understandable, but it's also dangerous.

Real leadership is the art of **starting where you are** with the people you actually have and helping them see themselves as part of something bigger.

Hiring as stewardship, not savior behavior.

The thread tying all of this together is simple:

Hiring is stewardship.

You are not rescuing yourself from overload by grabbing the first technically competent person who applies. You are making a choice that will shape the daily experience of everyone on your team, the trust others place in GIS, and the culture your future hires will walk into.

That's a big responsibility, but it doesn't have to be paralyzing. It just means slowing down enough to ask:

- "Is this person aligned with the journey we're on?"
- "Am I hiring them because they fit a story I've always told myself, or because they fit what this program truly needs?"
- "Am I willing to invest in their growth, not just squeeze output from them?"

In the next chapter, we'll talk about what happens **after** people join you:

How to build a team, not a hero factory. How to create conditions for trust, not quiet burnout.

How to make sure your people are growing, not just grinding. But before we go there, take a moment to honor this:

If you are even thinking about hiring differently, you are already leading differently.

Leader's Map in Practice - Chapter 10

Choose one or two questions and write down a real answer—ideally with names, not just abstract ideas.

1. Think about your GIS program 12 to 18 months from now. What will you need more of—deep technical build, relationship-building, data discipline, storytelling, training, or something else? How does that answer shape the **next role** you want to add to your team?
2. If you could design your ideal "complement"—someone whose strengths balance your limitations—what would they be like? What do they love doing that drains you? What do they see quickly that you tend to miss?
3. If you already have a team, think of one person you didn't hire. What is one conversation you could have with them this month that would help you better understand their strengths, aspirations, and frustrations?

CHAPTER 11:

Building a Team, Not a Hero Factory

IF YOU WANT TO know what kind of leader you've been, look at what your team feels they have to be in order to survive:

Heroes.

Martyrs.

Quiet workhorses.

Or a team of adults who trust one another and know where they're going.

Many GIS and IT leaders don't set out to build hero factories, but it happens accidentally.

You start out as the person who can do everything. You're fast, reliable, and you care, so people learn that if they come to you, things get done. When you finally get to hire, you look for people who remind you of...you.

They're smart. They can juggle. They'll "go the extra mile." At first, it feels amazing. At last, you're not alone.

Then, slowly, you notice the symptoms:

- Work is still overly dependent on a few people.
- Working nights and weekends are treated as "normal" for your high performers.
- Documentation and cross-training always lose out to urgent delivery.
- New people are quietly told, "We just figure it out here."

From the outside, it might look like a high-performing team. Inside, it's a collection of individual heroes holding up too much weight.

This chapter is about choosing something different.

What a hero factory feels like from the inside.

If you've ever been someone else's "go-to person," you already know what a hero factory feels like.

At first, it's flattering. You like being trusted. People bring you interesting problems. You get put on the important projects. Your name shows up in thank-you emails.

Over time, it starts to cost more than it gives. You can't take a real vacation without spending a week preparing and another week catching up. You wake up at 3 a.m., wondering what you forgot. You find yourself guarding knowledge, not because you're selfish, but because you are afraid that if you let go, the quality will drop. You start to resent the people who rely on you most.

If you are leading a team and any of your people are living like this, you don't have a sustainable system. You have a fragile one.

A leader in one of my conversations admitted, "I realized I had rebuilt my own burnout at scale. I had three people living the life I used to live. That's when I knew we had to change."

The hard part is that hero behavior is often rewarded. Heroes get praise, special assignments, and recognition. People who set reasonable boundaries can be seen as "less committed."

Changing that pattern is not about scolding heroes. It's about rewriting what "good" looks like on your team.

Redefining what "good" looks like.

On a hero factory team, "good" usually means taking on more, answering quickly, fixing problems alone, or saying yes.

On a healthy team, "good" looks different. It looks like:

- surfacing problems early instead of hiding them;
- asking for help before things break;
- documenting so others can succeed without you;
- protecting time for thinking and improvement, not just delivery;
- saying "not now" when capacity is full.

That shift doesn't happen by accident. You, as the leader, have to name it.

You do it in small ways. When someone admits a mistake early, you thank them for the transparency instead of punishing them for not catching it. When a team member shares a cleaned-up process or documentation, you call that out as real impact, not "extra." When someone is at capacity and says so, you treat that as responsible behavior, not a lack of dedication.

You also model it yourself.

If you answer emails at all hours, never take time off, and quietly fix everything alone, your team will assume that is the standard. No amount of slides about "self-care" will change that.

If you are willing to say, "I'm at capacity. I'll need to push this to next week," or, "I don't know, let's think it through together," you give them permission to be human too.

Designing roles so people can breathe.

Back in earlier chapters, we talked about seeing your program through a maturity lens and being honest about your capacity. Building a team, not a hero factory, also means designing roles with that honesty.

Instead of one vague role that says "do everything GIS," you begin to shape **zones of responsibility**: operations and support, data stewardship, applications and analytics, engagement and training, and architecture and integration.

In a small team, one person may hold more than one zone, but at least you've named them. Over time, as you grow, you can rebalance so that no one is carrying too many zones at once.

You also think about **backup**.

If one person is the only one who knows how to maintain a critical system, that's a risk, not a badge of honor. Cross-training doesn't have to be elaborate. Sometimes it starts as pairing on a task once a week, recording a short walkthrough, or maintaining a simple runbook for key processes.

This is not about replacing your experts. It's about making sure they are not trapped by their own value.

In one organization, a GIS lead realized that their most senior analyst had not taken a real break in years because "no one else knows how to run the updates." They made it a priority to train a second person, even though it slowed them down in the short term. The next time that analyst took two weeks off and everything kept working, it changed the way the whole team thought about resilience.

Creating psychological safety, not performance theater.

When we talk about psychological safety, it can sound abstract or "soft." In practice, however, it is one of the most practical ingredients of a functional team.

Psychological safety is not about everyone being nice all the time. It's about people feeling safe asking questions without being mocked, admitting when they don't know something, bringing up risks or concerns, and disagreeing respectfully.

On a team without that safety, people nod along in meetings and then quietly do what they were always going to do. Problems surface late, when resolving them is expensive. Mistakes are hidden, and feedback goes underground.

On a team with safety, you hear things like:

- "I'm not sure this will scale. Can we test it on a smaller group first?"

- "We've tried something similar before; here's what we learned."
- "I'm uncomfortable with this deadline unless we drop something else."

You don't get that by giving a speech about safety. You get it from your reactions. When someone voices a concern, do you listen, or do you steamroll them to protect your timeline? When someone gives you hard feedback, do you get defensive, or do you say, "Thank you. I needed to hear that." When something goes wrong, do you look for someone to blame, or do you ask, "What in our system made this likely?"

Your team is always taking notes, even when they're not saying so out loud.

Balancing standards and care.

You can have clear expectations about quality, follow-through, and professionalism, *and* recognize that your team is made of human beings with lives, families, and limits.

Stewardship doesn't mean lowering your standards. It means holding standards and care simultaneously, so your team becomes a place where people can do excellent work without leaving their humanity at the door.

One manager keeps a simple phrase taped near her monitor: "Remember the human." It is borrowed from an online community guideline, but it has become her leadership rule. Staff, stakeholders, vendors, and even difficult colleagues all have stories, griefs, and growth edges you cannot see.

This balance is especially important when someone is struggling.

If a team member starts dropping balls, the hero factory reaction is to dump more on the "reliable" people and quietly decide that the struggler "just isn't cut out for it." Resentment grows on all sides.

A stewardship response sounds more like:

> *"I'm noticing some things slipping, and that's not like you. What's going on? What do you need? What's in your way?"*

Sometimes the answer is skill: they need training or practice. Sometimes it's capacity: they're carrying too much. Sometimes it's personal: something hard is happening at home.

You still have to hold the line on what the role requires, but you can be clear *and* kind. You can say, "I believe you can do this, and we need to figure out how. Let's make a plan."

As you model that mix of standards and care, your team learns to treat one another the same way. Instead of gossiping when someone is off, they start to ask, "How can we help?"

That's when a group of individuals begins to feel like a team.

Recognizing and rewarding the right things.

What you recognize publicly becomes what your real values are.

If you only ever celebrate last-minute heroics—the person who saved the day with a midnight fix, the big sprint at the end of a project—you are teaching your team that chaos and sacrifice are how to earn praise.

If you also celebrate the person who quietly improved a process so you don't need last-minute sprints, the analyst who documented a workflow so others can run it, the team who delivered something on time *without* drama, or the colleague who mentored a newer staff member into confidence, you are telling a different story:

> *"We value sustainability as much as we value intensity."*

It doesn't have to be elaborate. Sometimes it's as simple as saying, in a meeting, "I want to acknowledge how you handled this. Because you documented and communicated clearly, we avoided a crisis."

Your team will listen more closely to those small moments than to any poster about values.

From team design to team alignment.

By now, you've thought about who you hire, how you shape roles, how you define "good," how you create safety, and how you balance standards and care.

The next question is: *aligned to what?*

Stewardship is not only about protecting each other from burnout, but also about aiming your collective energy at something that matters.

You are not building a healthy team for its own sake. You are building a healthy team so you can pursue a meaningful **North Star** together.

In the next chapter, we'll talk about **buy-in and opportunities to grow**: how to align your team with the vision, make space for their aspirations, and invite them to shape the path instead of just carrying out your plans.

Because the real power of a team is not in how well they follow you, but in how deeply they understand *why* you are all doing this work together.

Leader's Map in Practice - Chapter 11

Pick one or two prompts and answer them honestly. This is where theory meets your actual team.

1. Think of one person on your team who has been playing the hero. What has your leadership rewarded, directly or indirectly? What is one conversation you could have this month to move from heroics to healthier, shared ownership?
2. Look at your current roles. Where is risk concentrated in a single person's head or hands (a system, a dataset, a relationship)? What is one concrete step you can take in the next 60 days toward cross-training or documenting that work?
3. Reflect on the last time someone brought you bad news early. How did you respond? If you could replay that moment in a way that builds more safety, what would you do differently next time?

CHAPTER 12:

Buy-In and Opportunities to Grow—Aligning Your Team With the Vision

THERE IS A MOMENT many leaders remember with a mix of embarrassment and clarity.

You've spent weeks crafting a GIS vision, refining it with executives, and turning it into a slide that finally says what you've been carrying in your head for years.

You stand in front of your team, full of hope, and walk them through it, talking about North Stars, strategy, and impact. You finish, look up, and... see polite faces.

No one is hostile. No one storms out. But the energy in the room doesn't feel like "we're in." It feels like "we'll see."

You've just discovered a hard but important truth:

A vision is not shared just because you presented it.

Your team has its own history, its own scars, its own questions—will this be another initiative that burns us out? Does this mean even more work with the same number of people? Is this really for us, or is it for leadership slides? Where do I fit in this picture?

Aligning your team with the vision is not about "getting buy-in" like a campaign. It is about **building a shared story** where their growth and the program's growth are intertwined.

Start where they are, not where you wish they were.

When you have finally found language for your vision, it is tempting to think, "If I just explain it clearly, everyone will feel the same way."

That rarely happens.

Your team didn't live your whole journey. They didn't experience your conference epiphanies, your late-night thinking, your coaching sessions, or your informal chats with other leaders. They've been busy living their own story: delivering work, dealing with outages, navigating politics, and carrying their personal lives.

So the first step in alignment is not another presentation. It is **listening.**

Simple questions in 1:1s and small group conversations go a long way:

- "When you think about GIS here, what do you hope we could be in a year or two?"
- "What frustrates you most about how we're working today?"
- "What kind of work lights you up—the kind that makes the day fly by—and what kind of work makes you feel deflated by the end of the day?"
- "What worries you when you hear words like 'strategy' and 'change?'"

You may hear things that surprise you. Someone who seems disengaged might be deeply idealistic and tired of false starts. Someone who looks comfortable might be bored and hungry for more. Someone who avoids certain tasks might be carrying a quiet fear that they're not good enough for that kind of work.

Starting there doesn't mean you abandon your vision. It means you let reality into the room.

Translate the vision into "what this means for us."

Once you have listened, you can begin to connect your vision to their world.

It is one thing to say, "GIS will be a trusted information system that supports enterprise decisions." It is another to say in a small meeting:

> *"For us, that means fewer 'fire drills' and more time for work we can be proud of. It means when the council asks for a map, we aren't scrambling—we already know our data is trusted. It means more of you will be in rooms where decisions are actually made, not just sending maps in afterwards."*

You don't have to have every answer. In fact, it's better if you don't pretend to.

It can sound like:

- "This is the picture I see. Some parts are still fuzzy. I want us to fill in those pieces together."
- "Here's what I know this will mean in the next 6 to 12 months. Here's what I don't know yet."
- "There are things that will be hard. There are also real opportunities for us to grow."

When people can see themselves in the story, they stop treating the vision as something that is happening *to* them and start exploring how it could happen *with* them.

Invite them into the "how," not just the "what."

Alignment is not agreement on every detail. It is a shared commitment to a direction.

You own the responsibility for the **what** and **why** of the vision. But if you hold on too tightly to **how** everything must happen, you leave very little room for your team's creativity and ownership.

One of the most powerful ways to build buy-in is to ask:

> *"Given this North Star, how do you think we should move toward it in the next 90 days?"*

You might do that in a planning workshop where you share the organization's priorities, restate the GIS vision in plain language, and show the current capacity reality. Then ask the team, "What would be the most meaningful use of our energy this quarter?"

You will hear ideas you would not have come up with alone, as well as concerns and constraints you might have missed.

You still make the final calls, but those decisions now sit on top of a **shared conversation**, not a private spreadsheet.

Over time, as people see their ideas reflected in the plan, trust grows. They stop assuming that "leadership" is something done behind closed doors and begin to see themselves as part of the leadership fabric.

Growth is not a reward at the end. It's built into the journey.

"Opportunity to grow" should not mean, "When you've proven yourself for long enough, I'll let you do something interesting."

If that's the model, your most capable people will either burn out or leave.

Growth can and should be built into the way you pursue the vision from the start.

Once you have a roadmap, you can look at each initiative and ask:

- Who is ready to lead this with support?
- Who could co-lead, stretching into new skills?
- Who could shadow this work to learn how it is done?
- Who could own a small, clearly defined piece?

Growth doesn't always mean promotion or a new title. It can mean a chance to present to leadership, facilitate a cross-department meeting, design a small pilot program, or mentor a newer staff member.

You won't get this perfect. Sometimes you will stretch someone too far. Sometimes you will realize you kept someone in their comfort zone too long. But if your team can feel that you are *trying* to give them room to grow, that matters.

One leader described the shift this way:

> *"I stopped thinking of our roadmap as my to-do list for the team and started thinking of it as a development plan we were all living inside."*

Same work. Different posture.

Individual aspirations, shared direction.

Alignment does not mean everyone wants the same career path.

In your conversations, you will find people who want to deepen their technical expertise and be known as experts, move toward leadership and people management, focus on field enablement, or be the bridge between GIS and other departments—or are still figuring it out.

Your job is not to script their future; it's to be honest about what is and is not possible in your context, to look for intersections between their aspirations and the program's needs, and to help them see that even small steps count.

Maybe you can't promise someone a formal promotion, but you can give them a project that builds the skills they'll need, advocate for them when opportunities appear, and write a recommendation when they are ready to move on.

When people feel that their growth matters to you, they are much more likely to stay engaged in the hard, slow work of building the vision.

When people are not aligned.

Sometimes, even after listening, inviting input, and creating growth opportunities, you will realize that someone is not aligned with the direction.

They may be excellent technically but fundamentally opposed to where you're headed. They may want a stable, unchanging environment while you are leading through transformation. They may be attached to the old story of GIS in a way that keeps pulling the team backward.

This is one of the hardest parts of leadership.

You can't force someone to want what you want. Nor can you let one person's resistance define the ceiling for everyone else.

Honest conversations matter here:

> *"Here's where we're going. Here's what this role will require in the next year or two. Here's what I'm seeing in how you respond to that. How does this land with you?"*

Sometimes, that conversation unlocks something. Sometimes, it surfaces that this is no longer the right place for them.

Supporting someone in finding a better-fitting role elsewhere can also be an act of stewardship. In the long run, it is kinder than keeping them in a role where they are quietly unhappy and holding back the team.

From individual growth to a shared North Star.

As you do this work—listening, translating, inviting, stretching—something subtle happens.

The vision stops being "your" GIS vision and becomes **our** story: who we want to be as a team, what we want to be known for, how we want to treat one another, and the people we serve.

In the next chapter, we will make that explicit.

We will talk about **defining your team's North Star**—not just what GIS is for in the organization, but what this particular group of people wants to be together. That shared North Star becomes a compass for decisions, growth, and culture, especially when things get tough.

Leader's Map in Practice - Chapter 12

Choose one or two prompts and write real answers, with names and specifics where you can.

1. Think of a time you presented your vision to your team. If you replay that moment now, what questions or fears do you imagine were quietly sitting in the room? What is one conversation you could have now to bring those into the open?
2. Look at your current 90-day roadmap. For one initiative, write down who is leading, who could co-lead, and who could shadow to grow. What small adjustment would make that initiative more of a development opportunity for someone?
3. Name one person whose aspirations you are not completely clear on. What is one gentle question you could ask in your next 1:1 to understand what "growth" actually means for them?

CHAPTER 13:

Team Opportunity to Grow—Defining Your Team's North Star

YOU ALREADY HAVE A GIS North Star for the organization: a clear sense GIS's purpose, the decisions it must enable, and how it supports the wider vision.

This chapter is about something more intimate—who we want to be as a team.

Not in a slogan-on-the-wall way, but in a lived way: what it feels like to work here, what we're known for when people talk about us in other rooms, and what kind of opportunities we create for one another.

A team without its own North Star will still move. Work will get done. Tickets will be closed. Projects will ship.

But under pressure, you will default to habits like heroism, quiet resentment, and survival mode.

A team with a North Star has a way to ask, together:

> *"Given who we say we are, what do we do now?"*

That question changes everything.

Your team is not just a set of roles.

It is easy, especially when you are busy, to think of your team in terms of positions: the analyst, the developer, the data person, the field support person, the manager.

The org chart is necessary, but it is not the whole story.

Your team is also the way people talk to each other when they're stressed, the jokes that show up in meetings, the way new people are welcomed (or not), and the stories you tell about past wins and failures.

Defining a team's North Star is about becoming conscious of that story and shaping it on purpose.

It doesn't erase your differences. It gives them a shared container.

A simple conversation, not a branding exercise.

You don't need a consultant or a half-day offsite to start this work. You need a bit of time, some honesty, and a willingness to listen.

Here is one way it can look.

You gather your team—all of them, if possible, or in small groups if you're spread across locations. You set the context:

> *"We've been talking about our GIS vision for the organization. Today I want to talk about us—who we want to be as a team over the next few years. This is not about slogans. It's about how we want to work and grow together."*

Then ask a few questions and capture what you hear:

- "When you think about the best team you've ever been part of, here or elsewhere, what made it feel that way?"
- "If people in other departments described us in three words, what would you want those words to be?"
- "Imagine it's two years from now, and we are proud of who we've become as a team. What's true about how we work and how we treat each other?"

You will get a mix of practical and aspirational answers:

- "We have each other's backs."
- "We don't panic when something breaks."

- "We are known for telling the truth kindly."
- "People feel safe asking us 'stupid' questions."
- "We're always learning something new."
- "We're not afraid to say no when it's the right thing."

Your job is not to wordsmith in the moment; it's to notice patterns.

Later, you can distill those patterns into a simple statement or a few short lines, such as:

> *"We are a team that tells the truth kindly, shares knowledge freely, and makes decisions with the long-term in mind."*

Or:

> *"We want to be known as a kind, reliable, and curious partner that helps others make better decisions."*

It doesn't have to be perfect; it just has to be honest enough that people recognize themselves in it and aspirational enough that it pulls the team forward.

Connecting the team's North Star to daily choices.

A North Star only matters if you use it. You can start small. When you plan the next quarter, ask:

> *"If we are serious about being a team that shares knowledge freely, what does that mean for how we split this work? Who needs to learn what, and how will we make space for that?"*

When a stressful incident happens, you can ask in the debrief:

> *"Given our commitment to being kind and reliable, how did we do? What helped? What got in our way?"*

When someone is exhausted, you can ask:

> *"If we mean it when we say we make decisions with the long-term in mind, what needs to change about our load or our expectations?"*

Don't use the North Star as a weapon: "You said you wanted this, so now you must..." Use it as a shared compass: "We said this matters to us. How do we live that here?"

Over time, those small, repeated moments turn a sentence in a document into lived culture.

Space to grow inside the North Star.

One of the quiet gifts of a team's North Star is that it can make growth less scary.

When people know what kind of team this is, what behaviors are honored, and how decisions are made, they can take more risks.

The analyst who wants to present for the first time knows you value truth and kindness over perfection. The newer team member who wants to propose an idea knows curiosity is part of the team's DNA. The experienced staff member who is tired can say so without fearing they'll be judged as "not committed."

Your North Star can also guide how you respond when someone wants to move on.

If part of your story is "we help people grow," then someone leaving for a new opportunity isn't a betrayal but, in some sense, a fulfillment. You can be honest about the loss and still say, "We're proud you're taking this next step."

That does not erase the practical difficulty of hiring and training, but it keeps you from clinging to people in ways that harm both them and the team.

Teams of one still have a North Star.

If you are reading this as a team of one, you might feel like this chapter doesn't apply to you.

It does.

You are still a team, and your future team will live inside the habits and stories you are building now.

You can ask yourself:

- "What kind of GIS presence do I want to establish here?"

- "When people think of coming to me, what do I want them to feel?"
- "What kind of opportunities do I want to create for myself and, eventually, for others?"

Your answers might guide how you respond to requests, how you document your work, and how you speak up in meetings. They might even guide how you decide which opportunities to accept and which to decline.

When you eventually hire, you won't start from scratch. You'll already have a sense of who you want to be together.

From Team North Star to self-leadership.

By now, you have a clearer sense of who you want to be as a team, a language, however rough, for your shared North Star, and a way to connect daily decisions to that identity.

In the next part of the book, we will turn the lens fully inward.

Because none of this—vision, systems, stewardship, team growth—is sustainable if you treat yourself as a machine.

You are not just leading others; you are also leading yourself: through seasons of scarcity, through conflict, through your own ego, fear, pride, and purpose.

That is where we go next.

Leader's Map in Practice - Chapter 13

As always, choose one or two prompts that feel most alive.

1. Imagine someone in another department describing your team two years from now. What three words would you love to overhear? What would need to change between now and then for those words to be true?
2. If you gathered your team tomorrow and asked, "What kind of team do we want to be?" What do you think they would say? What would surprise you about that answer?
3. Write a first draft of your team's North Star in one or two sentences. Don't worry about perfect wording. How could you test this draft with your team in a way that invites their voices, not just their approval?

Part IV: Leading Yourself - Becoming the Leader You Seek

YOU CAN BUILD VISION, systems, roadmaps, teams, and governance.

The real test comes when all of that collides with your humanity.

The emails that arrive when you are already tired.

The stakeholder who publicly questions your work.

The staff member who is struggling and reminds you of an old version of yourself.

The night you look at your calendar and quietly think, "I cannot keep living like this."

Part IV is about that layer—the self-leadership that makes the rest of the book possible.

You are not just a GIS leader. You are a human with a history, a nervous system, a heart, and a future you once felt but could not quite see. How you treat yourself, what stories you believe about your worth, and how you handle fear and pride will shape everything you build.

In these final chapters, we will look at:

- How to manage your own time and energy, so your future self has a chance.
- How to listen—to others, to the system, and to yourself.
- How to walk into conflict instead of circling around it for months.
- How to notice when your ego is driving and choose purpose instead.
- How to move, over time, from scarcity to real influence.

This is the inner map.

Because in the end, the most powerful GIS tool you will ever work with is not the technology.

It's the leader you are becoming.

CHAPTER 14:

Leading Yourself: Forward-Looking, Dependable, Courageous

WHEN I ASK LEADERS what they most want to be known for, their answers sound similar:

"I want people to trust me."
"I want them to know I have their back."
"I want to be someone who sees around corners, not just reacts."

In my own leadership journey, three qualities kept showing up in the people I admired most—they were forward-looking, dependable, and courageous.

None of them were perfect. They forgot things, they had bad days, and they made mistakes. But over time, these three traits were like a spine everything else could rest on.

This chapter is about practicing those traits with yourself first.

Forward-looking—giving your future self a chance.

Forward-looking does not mean predicting everything. It means acting as if your future self matters at least as much as everyone else's urgent needs.

When you are not forward-looking, your days fill themselves—you accept every meeting because it is on the calendar; you say "yes" to every request because you do not want to disappoint; you tell yourself you'll work on strategy "when things slow down"—and they never do.

One of the simplest tools I've used to shift this is the classic **urgent–important matrix**, attributed to President Dwight D. Eisenhower, who famously said:

> *"What is important is seldom urgent,*
> *and what is urgent is seldom important."*
> —Dwight D. Eisenhower

Think of your work as falling into four quadrants:

1. **Important and urgent**—fires you really do need to put out;
2. **Important but not urgent**—work that shapes the future;
3. **Not important but urgent**—noise that demands attention;
4. **Not important and not urgent**—distraction.

Most GIS leaders I talk to live in **Quadrants 1 and 3**: important/urgent and not-important/urgent. They're constantly reacting.

Let's give each quadrant a face in your world.

Quadrant 1—Important and urgent.

These are real emergencies: a critical web map is down during a council meeting; a field app is broken, and crews are idle; or a data error has made it into a public-facing product.

You can't ignore these; it's part of service delivery. But if every day feels like only this, you don't have a leadership job; you have a permanent crisis-management job.

Quadrant 2—Important but not urgent.

This is the quadrant leaders starve for and then wonder why nothing changes.

Quadrant 2 includes drafting and refining your 90-day roadmap; running a governance meeting that aligns priorities before they become conflicts; building relationships with key executives before you "need" something; mentoring a team member; or documenting a process so you don't have to be the bottleneck next time.

Nothing explodes if you delay these by a week, and that's the danger. They're easy to postpone, and yet they are exactly where your **future influence** is built.

Quadrant 3—Not important but urgent.

This quadrant is sneaky. It feels like work, but it's often other people's noise: meetings where you don't truly need to be there, "Can you just..." requests that don't align with any roadmap, reports no one really reads, and work done because "we've always done it."

Quadrant 3 drains the time you need for Quadrant 2.

Quadrant 4—Not important and not urgent.

This one you already know: doom-scrolling, checking email for the twenty-seventh time, rearranging layers in a map because you're avoiding a harder task.

Everyone does some of this. The goal isn't perfection; it's awareness.

Forward-looking leadership is the practice of shifting more of your energy into Quadrant 2 on purpose.

That might look like blocking two 90-minute "strategy" sessions on your calendar each week and treating them as sacred, cancelling or declining one recurring meeting that lives in Quadrant 3, or setting aside an hour on Fridays to review your 90-day roadmap and adjust it based on what you learned.

If someone looked at your calendar, could they tell what your future self cares about?

Being forward-looking is not about heroic discipline, but about designing your time so your leadership has a chance to exist.

Dependable—treating your time as a promise.

Dependable does not mean "never dropping a ball." It means people can trust what your yes and your no actually mean.

If you say yes to everything, you are not being generous; you are **making promises you cannot keep**.

Leaders I interviewed who were known as steady and trustworthy had a few things in common:

- They were honest about capacity. Instead of "We'll see what we can do," they said, "Here's what we can do with the time and people we have."
- They pushed for prioritization, saying, "If everything is priority one, nothing is."

- When they did miss, they owned it early: "I'm going to be late on this. Here's why, here's what I'm doing about it, and here's what we can adjust."

Dependability starts with how you treat **your own calendar**.

You could think of every slot as a small contract. When you schedule time for strategy, 1:1s, or deep work and then repeatedly give that time away to the loudest urgent thing, you are breaking contracts with yourself and with the people who depend on you long-term.

This is where the Eisenhower model comes back. If you want to be dependable, you need **enough Quadrant 2 time** to keep your bigger promises—the roadmap, the governance cadence, the growth of your team.

Dependability also lives in how you respond when capacity changes—when a key staff member is out and you proactively reset expectations with stakeholders, or when a project turns out to be more complex than you thought and you tell the truth rather than silently hoping it will somehow work out.

People don't need you to be superhuman. They need to know that when you say, "I'll do this," it rests on real thought, not wishful thinking.

Courageous—shifting from path to quest.

Courage in leadership is not just about big, dramatic decisions. It's about the quieter choice to **step into the unknown** on purpose.

A lot of leadership training focuses on what my coach and teacher, **Robert Ellis**, would call the **path**: going from A to B. You're at A, you want a GIS roadmap (B); you plan tasks, milestones, and deliverables. Path is important. It's project management, execution, and delivery.

But Robert makes a crucial distinction: if you only live in the path, you limit B to what you already know.

A **quest** is different.

You still start at A. You still have a B in mind—the kind of leader you want to be, the kind of GIS program you want to build, the kind of culture you want to create. But you hold B lightly. The real aim is **B-prime**—a version of success you can't fully picture yet when you start.

Quest leadership has a few key ingredients:

- You accept that uncertainty is not a bug; it's the space where better possibilities live.
- You know you'll have to try things, learn, and adjust.
- You keep asking better questions:
    - "What is truly missing here?"
    - "What are we longing for that this system doesn't yet provide?"
    - "What kind of work is trying to be done through us that we're not allowing yet?"

For a GIS leader, a quest might look like deciding to move from "map service" to "decision support partner" without knowing exactly how that will play out, only that staying where you are is no longer acceptable; or committing to build a healthier team culture in a place where burnout and cynicism have been normal for years; or choosing to speak in outcomes to executives, even though you feel safer talking about layers and services.

On a path, you mostly manage tasks. On a quest, you also **change yourself** as you go.

That requires courage, because you cannot control all of it. You will make bets that don't pay off immediately. You will need to say, "We tried this, we learned this, and here's how we're adapting."

Forward-looking dependability keeps your quest from being a fantasy. They give it bones. The quest mindset keeps your leadership from becoming a to-do list you could hand to anyone.

Thresholds—the invisible lines you must cross.

Even when you understand the Eisenhower quadrants and the idea of a quest, you'll still hit internal walls.

Robert calls these **thresholds**—invisible lines where your system wants **you** to stop, the point where something feels too silly, too uncertain, too vulnerable, or too "not me."

He identifies several thresholds that show up over and over in leaders: **silliness, knowledge, trust, sanity, and love**.

Let's look at each in the context of GIS leadership.

The silliness threshold.

This is the edge where you feel foolish when trying a new facilitation exercise in a governance meeting, asking executives to draw their ideal future on sticky notes, or running a "show and tell" of failures in your team.

Part of you says, "This is dumb. Serious people don't do this."

On the other side of the silliness threshold are **play, creativity, and connection.** Many of your most effective experiments will feel a little silly at first. Crossing this threshold might mean **letting yourself try** anyway, in low-risk ways.

The knowledge threshold.

This is the edge where you admit you don't know. It might look like telling a stakeholder, "I'm not sure yet; here's what we need to find out," asking a team member to explain a technology they know better than you, or enrolling in a training that exposes your gaps.

If your identity has been built on being the expert, the knowledge threshold can feel threatening. But refusing to cross it traps you in old tools and prevents your team from growing.

On the other side is **real learning and shared ownership.**

The trust threshold.

This is about letting go of control: delegating a high-visibility task to someone else, letting a team member present to executives while you sit in the back row, or sharing early, messy drafts of a strategy instead of waiting until it is "perfect."

Your ego might say, "If I don't tightly manage this, things will go badly and reflect on me."

Crossing the trust threshold is how you **build capacity** rather than become the bottleneck. It's also how you show your team you truly believe in them.

The sanity threshold.

This is your tolerance for ambiguity and change. It might look like working on a roadmap while the organization is in flux, starting a pilot when you can't guarantee the outcome, or saying yes to a strategic opportunity even though not all the details are nailed down.

At the sanity threshold, you might feel anxious, scattered, and tempted to run back to busywork because it feels controllable.

Crossing it in small steps is how you grow your ability to **lead through uncertainty instead of only in stability**.

The love threshold.

This is the deepest and often the most tender threshold.

It's about how willing you are to be seen when admitting you made a call that wasn't aligned with your values and correcting it; telling your team, "I'm learning this too; here's where I've been scared or stuck," or—my favorite—letting yourself receive appreciation instead of deflecting it with a joke.

If, somewhere inside, you carry the belief "if people really see me, they'll reject me," the love threshold will keep you armored.

Crossing it is not about oversharing. It's about letting yourself be a **whole human** in your leadership—not just a title with a to-do list.

On the other side is deeper **trust**, with yourself and with the people you serve.

Forward-looking, dependable, courageous leadership isn't a personality type you either have or don't. It's a series of small, repeated choices: to design your time instead of letting urgency own you, to treat your commitments as promises (not wishes), to approach your work as a quest (not just a path), and to notice your thresholds and step through them, gently, one at a time.

In the next chapter, we'll focus on one of the most practical expressions of all this inner work: how you **listen**—to others, to the system, and to yourself.

Leader's Map in Practice - Chapter 14

Pick one or two prompts and write real answers, with specifics.

1. Take one day from your calendar and label each block as Quadrant 1, 2, 3, or 4. What do you notice? What is one concrete change you can make this month to create more Quadrant 2 time?

2. Think of a situation where you held back because you "didn't want to look silly," "should already know this," or "didn't trust others to handle it." Which threshold was that? What would a small, safe experiment in crossing that threshold look like?

3. If you think of your next 6 to 12 months as a quest, not just a path, what is B (what you think you want) and what might B-prime be (a version of success you can't fully define yet but can feel)? What is one good question that could guide you on that quest?

CHAPTER 15:

Listening Beyond the Map: Active Listening as a Leadership Superpower

WHEN YOU ASK PEOPLE what made a past leader unforgettable, they rarely say, "They had the best dashboards."

They say things like: "They actually listened to me," "They made me feel seen," and "They heard what I was trying to say, not just what came out of my mouth."

In GIS, it is easy to think your job is to listen to data—layers, attributes, patterns. But leadership demands another kind of listening:to people, to yourself, and to the system around you.

Active listening is not optional. It is how you find the real problems to solve, the real fears behind resistance, and the real opportunities for growth.

Listening as hosting, not performing.

One of the practices I learned from **Robert Ellis** is called "Be the Host." The idea is simple: in any meaningful conversation, you can choose to be the host or the guest.

As the guest, you try to impress, you worry about how you look, and focus on saying the right thing so you will be invited back.

As the host, your attention is on your guests—their comfort, their experience, what will serve them. You are still fully yourself, but you are not the main character.

Bringing this into leadership shifted how I approached meetings and 1:1s. Instead of walking in wondering, "How do I get through my agenda?" I started asking, "What is really happening for the person in front of me? How can I make it safe enough for the truth to show up?"

Several leaders in the interviews did this intuitively. One described walking into a monthly check-in with an underperforming staff member and starting with, "Before we talk about tasks, how are you really doing?" Another talked about spending the first ten minutes of a stakeholder meeting just asking about their world—pressures, goals, worries—before ever mentioning GIS.

Hosting does not mean rescuing but listening in a way that invites the full story, not just the polished version.

SACRED PANDA—listening in a structured way.

Listening can feel abstract until you give it a shape. One of my favorite shapes is a pattern Robert calls SACRED PANDA. Let's talk about the SACRED part first:

- Successes
- Appreciations
- Challenges
- Requests
- Experiments
- Dreams

We used this in leadership groups and 1:1s.

You might ask a team member:

- "What **successes** have you had since we last spoke—wins that you are proud of, big or small?"
- "Is there anyone you want to **appreciate**—someone who helped you or modeled something well?"

- "What **challenges** are you facing—where do you feel stuck or frustrated?"
- "Do you have any **requests** for the team or me?"
- "Is there anything you want to **experiment** with in the next couple of weeks?"
- "Looking a bit further ahead, what are you **dreaming** about for your work or career?"

The order matters. When you start with success and appreciation, you remind people they are more than their problems. By the time you get to challenges and requests, there is already some trust in the room. When you ask about experiments and dreams, you signal that their growth matters, not just their output.

You do not have to use the SACRED acronym explicitly with your team. You can hold it in your own mind as a checklist to guide your listening.

Triple-loop learning—listening for the deeper pattern.

Another tool from Coaching from Essence that changed how I debriefed experiences is triple-loop learning.

Single-loop learning asks: "What happened?" Double-loop learning asks: "How did we approach it—what process or assumptions were we using?" Triple-loop learning asks: "Who were we being—what identity or story were we living from?"

Imagine a project that went badly—a blown deadline, a frustrated stakeholder, a tired team.

Single-loop learning might focus on tasks: we underestimated the effort, we had a vendor issue, or we had too many dependencies.

Double-loop learning might look at the process: we did not involve the stakeholder early enough, we skipped testing, or we had no clear owner.

Triple-loop learning goes deeper: we were being heroes instead of stewards, avoidant instead of honest, or order takers instead of partners.

Listening at that third level can be uncomfortable. It asks you to see how your own identity stories are shaping your decisions, but that is where the most powerful change lives.

Several leaders described doing their own version of this after tough seasons. One GIS manager talked about realizing, after yet another "fire drill" project, that she was addicted to being the firefighter. Another noticed that he always said yes to senior leaders, even when it hurt his team, because he was still operating from a story that his worth depended on being liked by people above him.

Triple-loop listening lets you notice those patterns without shame and gives you language to say, "We were being X. Next time, I want us to practice being Y."

Listening to yourself.

Active listening is not just outward; it also means listening inward. The tightness in your chest before a meeting with a certain person, the sinking feeling when a new project is announced and you know the timeline is a fantasy, the small voice that says, "This is not sustainable" when you open your laptop at 10:00 p.m. again.

You can override those signals in the name of being a team player—many of us have—but over time, ignoring them leads to burnout, resentment, or quiet disengagement.

Listening to yourself does not mean obeying every feeling, but It means taking your signals seriously enough to explore them.

You might ask:

- "What is this tension trying to tell me about my boundaries, my values, my fear?"
- "If a staff member told me they felt this way, what would I advise them to do?"
- "Where am I pretending things are fine so I do not have to have a harder conversation?"

In my own life, there was a time when my body knew before my mind that something had to change. I was exhausted, anxious, and running on autopilot. It took heartbreak and coaching to finally listen to what had been there all along: a deep desire to lead in a way that was aligned with who I am, not just what I could produce.

Listening inward is not self-indulgence. It is part of being a trustworthy leader. People can feel when you are at war with yourself.

Leader's Map in Practice - Chapter 15

Choose one or two.

1. Think of a recent conversation where you realized only afterward that you had been rehearsing your response instead of truly listening. What might have changed if you had approached the conversation as the host rather than the guest?
2. Try a SACRED-style check-in—with yourself, a peer, or a team member. What surprised you when you listened for successes, challenges, and dreams, instead of just status updates?
3. Reflect on a recent mistake or conflict. Answer the triple-loop questions: What happened? How did we approach it? Who were we being? What might we choose to be next time?

CHAPTER 16:

Conflict and Difficult Conversations: Turning Tension Into Alignment

MOST LEADERS DO NOT need more training on how to avoid conflict. They are already experts.

We delay feedback until performance reviews. We complain to peers instead of speaking to the person directly. We escalate to HR or our own boss only when things are on fire.

For a long time, I treated conflict as a sign that something was wrong with me or the relationship. If I were a better leader, I thought, these situations would not exist, so I tried to smooth them over, work around them, fix things quietly.

All that did was let small fractures grow in the dark.

This chapter is about seeing conflict for what it really is: a normal part of human systems and an opportunity to move closer to alignment if you are willing to walk toward it.

The cost of not having the conversation.

Research from places like Harvard Business Review and the Society for Human Resource Management (SHRM) has been consistent: avoiding hard conversations costs organizations real money and trust. Employees spend hours ruminating, productivity drops, and good people leave because tensions never get addressed.

In GIS and IT, the costs are often quieter: a team member quietly disengages because they never hear the truth about their performance; a stakeholder loses trust because issues are never clearly named; or a leader becomes the "nice one" but not the person people go to when things really matter.

In one of my own roles, I watched a performance problem play out over months. A staff member was consistently late with deliverables, defensive about feedback, and difficult in meetings. I tried gentle nudges, extra support, and rewriting tasks to make things easier. I did everything except sit down for a clear, direct conversation about expectations, impact, and what needed to change.

By the time I finally had that conversation, the rest of the team was frustrated, my own credibility had been quietly eroding, and the staff member felt blindsided. We got through it, but it was much harder than it needed to be.

Avoiding conflict doesn't protect people. It leaves them alone with problems they cannot see or fix.

What makes hard conversations so hard?

If you have ever stared at your screen, avoiding sending an email or felt your stomach clench walking toward a meeting, you are not weak. You are human.

Hard conversations poke at thresholds—especially the love threshold Robert describes—raising questions like:

- "If I tell the truth, will this person still like me?"
- "If I admit my part in this, will I be seen as a failure?"
- "If I set a boundary, will I be punished?"

We also carry stories from past experiences. Perhaps you grew up in a home where conflict meant shouting, withdrawal, or worse. Perhaps you worked for someone who weaponized feedback. Those histories sit in the room with you.

Acknowledging that this is hard does not mean you get to skip it, but rather that you approach it with compassion—for yourself and the other person.

A simple structure for difficult conversations.

You do not need a perfect script, but you do need a structure.

Here is a simple one adapted from leadership research and my own practice:

1. **Prepare yourself, not just your notes.**
    - o Get clear on the facts: what actually happened, what you observed, and what the impact is.
    - o Notice your own story: are you going in already convinced they are lazy, disrespectful, or impossible?
    - o Decide your intention: alignment, clarity, support, or—if necessary—a clean ending, not punishment.
2. **Describe what happened.**
    - o Use concrete examples, not general accusations; for example:

> *"In the last three weeks, these two deadlines were missed, and I heard from two stakeholders who did not receive updates."*

3. **Share impact and invite their view.**

"Here is how this is affecting the team and our commitments."

   - Then ask, *"How do you see it? What is your experience of this?"*

4. **Listen, really listen.**

   This is where your Chapter 15 skills come in.

   - You might hear information you did not have—personal struggles, unclear expectations, obstacles you missed.
   - Listening does not erase the impact; it gives you the full picture.

5. **Clarify expectations and options.**

"Here is what needs to change, and by when."

   - If appropriate: *"Here is what support I can offer. Here is what I cannot do."*

6. **Agree on next steps and follow up.**
   - Summarize: *"Here is what we are both committing to."*
   - Set a time to check in.

This structure can also be used upwards and sideways, with some adaptation. You can say to a stakeholder,

"When priorities change mid-sprint without a conversation, here is the impact on the team and our ability to deliver. Can we talk about how to handle this differently?"

Practicing on small things.

You do not begin with the hardest conversation of your life.

You practice on smaller ones—asking a peer to stop interrupting you and others in meetings, telling a stakeholder, "I am at capacity; I can do X now or Y later, but not both," or asking your own boss for clearer priorities.

The more you experience yourself surviving these conversations, the less power your fear holds.

One GIS leader told me that they made a private agreement with themselves: if something bothered them for more than 24 hours, they would address it within a week. At first, that felt terrifying. Over time, it became a relief—a way to prevent resentment from building.

Leadership is not the absence of tension but the willingness to walk into it thoughtfully.

Leader's Map in Practice - Chapter 16

Choose one.

1. Think of a conversation you have been avoiding. Using the simple structure above, sketch how you might open it and what your intention would be.
2. Recall a past conflict you handled poorly or avoided. What was the cost to you and others? With the tools you have now, what might you do differently?
3. Write down one belief that makes hard conversations feel dangerous for you—for example, "If I say this, they will leave" or "Good leaders do not upset people." Is that belief always true? What might be a more accurate one?

CHAPTER 17:

Pride or Purpose: The Dark Side of Leadership

LEADERSHIP WILL SHOW YOU your best qualities. It will also expose your most fragile places.

The first time someone introduces you as a director, manager, or "the person who made this possible," it feels good. Your work is visible. Your effort is paying off.

The danger sneaks in quietly, underneath.

You start to believe that your value is tied to always having the answer. You start to protect your image instead of the truth. You start to treat dissent as disrespect and feedback as a threat.

This chapter is not about shaming you, but about naming the patterns so you can choose something different.

When ego takes the wheel.

There was a time in my life when my identity was deeply tied to being seen as strong and capable. At work, that meant taking on more than was healthy, hiding my doubts, and avoiding any vulnerability that might crack the image.

In my personal life, the same pattern played out in a relationship that became a mirror I did not want but desperately needed.

Without going into specifics, here is what matters for this chapter: I found myself in a situation where I was working hard to be chosen, to be the one who was "enough" for someone else's complicated world. When things did not go the way I wanted, my pride was devastated, and I felt rejected, unworthy, and foolish.

For a while, I turned that pain inward, thinking, *I must have done something wrong, I must not be lovable, I should have seen it coming*. Then, with help from therapy and coaching, I began to see the deeper pattern.

I had been using the relationship the same way I had used work—as proof that I was enough, that I was valuable, that I deserved my place.

That is the dark side of leadership: using external roles and relationships to patch internal holes.

Several leaders I spoke with recognized themselves in this.

One admitted that they could not handle not being the smartest person in the room and would subtly undermine staff who had ideas they had not thought of. Another confessed to avoiding promoting a talented team member because they did not want to be outshone.

Others talked candidly about times when their ego made them defensive with executives and how much more effective they became when they learned to separate feedback on their work from their worth as a person.

Pride as a shield for fear.

Robert's thresholds model helped me see pride differently. Pride, in this sense, is often a shield against a perceived threat to one's love threshold.

If, deep down, you fear that you are unworthy of love or respect unless you are perfect, your ego will do whatever it can to protect that image.

It might show up as dismissing others' contributions to stay on top, explaining away mistakes instead of owning them, reacting with anger when someone points out a blind spot, or clinging to old ways with the justification that "this is how we have always done it."

Underneath, there is usually a scared part terrified of being seen as inadequate.

Purpose is something else entirely.

Purpose cares less about your image and more about your impact on others. It can admit, "I do not know," because what matters is finding a good answer, not being the one who has it. It can apologize, change course, or step aside when someone else should lead.

The shift from pride to purpose is not a one-time decision but a lifetime of small choices.

Catching yourself in the mirror.

You cannot uproot the ego you refuse to see.

One practical way to begin is to notice your reactions in certain situations: when someone else gets credit you feel you deserve, when a team member brings you bad news, when a new person joins with strong skills, or when a peer challenges your idea in front of others.

Do you feel a tightness, a rush of defensiveness, or a desire to explain, correct, or withdraw?

That is ego knocking.

In those moments, you can give yourself a tiny pause—one breath—and say internally: "This is pride talking. What would purpose do here?"

Purpose might say, out loud, "Thank you for catching that," or "Tell me more," or "You make a good point—I had not thought of it that way."

You will not always get it right, but the more you practice, the easier it becomes to notice when you are slipping into old patterns.

Choosing purpose when things hurt.

There will be times when leadership breaks your heart: a project you poured yourself into gets cancelled; a boss you trusted lets you down; a team member you invested in leaves unexpectedly; a board chooses someone else for a role you wanted.

In those moments, pride will whisper, "How dare they, after everything you have done. You will show them."

Purpose asks a different set of questions:

- "What matters most in this situation?"
- "What kind of leader do I want to be in how I respond?"
- "What would my daughters, or the people I mentor, learn from watching me handle this?"

After my own season of heartbreak, I remember standing outside on a cold day, feeling the wind on my face, hand on my heart, and asking myself, "How do I want to come home to myself through this? Who do I want to be on the other side?"

That question did not erase the pain, but it did shift my focus from proving my worth to living my values.

You cannot control how others behave. You can choose whether your leadership will be driven by pride or anchored in purpose.

Leader's Map in Practice - Chapter 17

Choose one.

1. Recall a recent moment when you felt defensive, jealous, or overly attached to being right. What might have been threatened in you? If you could rewind that moment and respond from purpose, what would you do differently?
2. Think of a disappointment in your career—a role you did not get, a project that failed, a leader who let you down. What did that experience reveal about your identity stories? What did it invite you to grow into?
3. Write a short paragraph that starts with, "When things do not go my way, the kind of leader I want to be is someone who..." Then list three concrete behaviors that embody that leader.

CHAPTER 18:

Embodying Leadership—Moving From Scarcity to Influence

BY NOW, YOU HAVE seen leadership from many angles.

Identity and vision. Systems and strategy. Teams and stewardship. Self-management, listening, conflict, ego, mentoring.

This final chapter is not about adding more content, but about naming the shift underneath all: the shift from scarcity to influence.

Scarcity is the story that there is not enough—not enough budget, time, authority, talent, love, support, or space for you.

Influence is not the same as power. It is the capacity to shape outcomes because people trust you, your systems, and your intentions.

The leaders who left the deepest mark on me were not always the ones with the biggest titles.

They were the ones who had quietly moved from scarcity to influence in how they lived.

Scarcity sounds like this.

Scarcity is quick. It lives in the amygdala, the survival brain. It sounds like:

- "I am just a GIS person."
- "If I say no, they'll go around me."
- "If I don't do it myself, it won't be done right."
- "If someone else shines, there will be less for me."
- "When things slow down, then I'll focus on leadership."

Some of these sentiments may even have felt true at times. Many people I interviewed started in organizations where GIS was misunderstood or underfunded, and they had to fight for every inch.

The danger is when those old survival stories keep running long after the context has changed, or long after you have outgrown them.

Scarcity makes you chase seats at tables instead of asking what you bring to the table. It makes you hoard work instead of sharing it. It keeps you too tired to see new options.

Influence looks different.

Influence starts with internal alignment.

You know who you are as a leader. You know what your work is for. You have enough clarity and systems to say yes and no with integrity.

From there, influence looks like speaking in the language of outcomes instead of features, asking better questions instead of trying to impress, designing one thoughtful win that changes how others see GIS, or creating containers where other people can grow into their own leadership.

In the interviews, examples of this were everywhere.

One leader described how, instead of begging for more budget, they brought three simple slides to their executive team: a clear North Star statement, a 90-day roadmap tied to organizational goals, and a snapshot of current capacity versus demand. They did not posture but told the truth and asked for specific decisions. The result was more support than they had expected.

Another talked about mentoring analysts across departments on basic data literacy, not because it was in their job description, but because it raised the overall decision-making quality. That investment paid off when those analysts grew into champions who pulled GIS into strategic conversations.

Influence is not louder; it is clearer.

From path to quest.

You could read this book as a path playbook: do these things, build these systems, have these conversations. There is value in that.

Underneath, though, this has always been about a quest.

The quest is not "become a GIS director" or "be invited to present at a conference," though those things may happen.

The quest is becoming the kind of leader your younger self needed, the kind of leader your community deserves, the kind of leader your children, if you have them, can look at and say, "So that is possible."

That quest does not end when the book ends. It will keep inviting you into deeper honesty, clearer boundaries, kinder courage.

You will face seasons of loss and seasons of joy. You will forget what you know and remember again. You will outgrow roles and stories. If you are lucky, you will be surprised by how much more influence you have when you stop chasing it and start serving from your essence.

The future you felt but could not yet see.

When I crossed that ocean years ago with a geography degree in my suitcase and imperfect English on my tongue, I had a feeling more than a plan. I knew I loved maps and wanted a bigger life. I did not know it would lead here—to GIS leadership, to trainings, to interviews with people around the world, to this book.

You may be standing at your own shoreline right now.

Maybe you are a team of one, buried in tickets, wondering if leadership is really for you. Maybe you are in the middle of your career, feeling stuck between operational work and the strategic seat you know you could hold. Maybe you are already in a senior role and secretly afraid that people will discover you are making it up as you go.

Wherever you are, you are not alone.

You have a community of people—some whose names you have read in these pages, many others you have not met yet—who are doing this work alongside you, mentors you can seek out, and people coming up behind you who will be shaped by how you lead.

You have, most importantly, your own essence—the part of you that knows what matters when all the titles and projects fall away.

The question now is simple and not easy:

> *"Given who I am, what I know, and whom I serve, what is the next courageous step I am willing to take?"*

You do not have to have everything figured out. You do not have to be ready.

You only have to begin—again, and again.

Leader's Map in Practice - Chapter 19

Pick one.

1. Write down three scarcity sentences you catch yourself thinking about your work or career. Next to each, write one influence sentence you would rather practice. For example, "I am just the GIS person" becomes "I run an information service that helps this organization make better decisions."
2. Imagine your younger self—the version of you who was just starting in GIS. What would they be most proud of in how you lead today? What would they ask you to keep fighting for?
3. Think of one concrete action you can take in the next 30 days that aligns with influence rather than scarcity—a conversation, a boundary, an experiment, or an invitation.

Acknowledgement

OMG, I CAN'T BELIEVE this book actually happened.

For a long time, it lived in sticky notes, late-night journal entries, and quiet promises to myself. Seeing it in the world feels both surreal and deeply right. If you are reading these words, it is because many people loved me, challenged me, and stood beside me long before I felt "ready."

First, to my daughters, **Tiffany and Gray**. Thank you for loving me in all my unfinished and complicated versions. You have watched me work late, question myself, start over, and keep going when it would have been easier to quit. You reminded me to laugh, to rest, and to keep my heart open. This book is as much for you as it is for the GIS community. When you look at these pages, I hope you see not just my career, but a woman who chose to live with courage and curiosity.

To my **parents**, who taught me how to work hard, how to keep my word, and how to start again in a new place with an accent and a suitcase, your sacrifices made it possible for me to cross an ocean, earn a degree, and build a life I could barely imagine when I was young. Dziękuję. Your example is woven through every chapter.

To my friends, especially **Tess** and **Sharyll,** and extended family, especially my aunt, **Ela**—thank you for cheering me on when this was still "just an idea," for the messages, the calls, the "how's the book going" nudges, the coffee breaks where we talked about everything and nothing, and for having faith that I would actually finish.

To my mentor, **Trevor**, who has been quietly pushing me into leadership for years. You often saw the next version of me before I did. You handed me meetings, presentations, and responsibilities that felt a size too big, and then stood close enough that I could borrow your courage until I found my own. Your wisdom is not always comfortable, but it has always been honest and deeply rooted in care. Thank you for trusting me before I trusted myself.

To my coach, **Robert**, founder and author of Coaching from Essence, who changed my life and whose fingerprints are all over the inner tools in this book. Robert lived and taught with stage IV prostate cancer, and on New Year's Day 2024, his body finally let go.

One of his guiding lines was: "*Cancer is my teacher. Love is the lesson.*" He wrote often about how he believed in us—that whatever we were longing for was possible, that our essence had to have its way in the world if we were ever going to be truly happy. He kept reminding us not to let fear run our lives, but to create from essence, abundance, service, and trust. He gave away his book and videos as a gift because he wanted as many people as possible to have access to that way of living.

When I met Robert, I was still wrestling with my own darkness—old stories about worth, fear of not being enough, parts of myself I didn't quite know how to love. His coaching and his work did not erase that darkness. Instead, he helped me see that it was part of the terrain, not a disqualification. He gave me the strength to love myself, not despite the darkness, but including it. To choose leadership that is grounded in essence and love, not in proving and scarcity.

Robert often ended his notes with three simple words: "I love you." This book, and the leaders it is meant to serve, are one small way I get to say back: Robert, I love you too. Thank you for everything you poured into me and into this work.

To **Paul**, for your obsession with strategy and maturity, and for showing up as both a colleague and a friend. Your work on geospatial roadmaps and our many conversations about the "why" behind them have had a direct impact on how I think about GIS as an enterprise system. Thank you for letting me learn alongside you.

To my publisher and the team at Epic Author, especially **Trevor, Len, and Mark**, thank you for believing this book belonged in the world and for helping shape a messy pile of ideas into something coherent and useful. Your guidance, patience, and structure made the process far less lonely.

To all the **GIS professionals and leaders who contributed their stories**—whether they hold impressive titles or lead quietly without one—thank you for opening your calendars and your hearts. You trusted me with your frustrations, your experiments, your failures, and your wins. Many of you are building programs in environments that do not yet fully understand your value, and you are doing the brave work of leading anyway. Your words are woven throughout this book, sometimes with your names and organizations, sometimes anonymized, but always with deep respect.

A special thank you to those who are often overlooked in leadership conversations: the **teams of one**, the analysts and specialists who carry an entire GIS "department" on their shoulders. Your reality is one of the reasons this book exists. I wanted you to see yourselves here, not as an afterthought, but as leaders in your own right.

To my **clients**, **colleagues**, and **peers** in the GIS and IT community who let me experiment in real life—at work, in workshops, leadership trainings, and strategy sessions—thank you. The frameworks and stories in these chapters were tested in your conference rooms, on your whiteboards, and in your daily work. You allowed me to translate theory into practice.

To the **communities** that have welcomed and amplified my work—Women in GIS, local and regional GIS councils, conference organizers, and countless online groups—thank you for giving me stages, microphones, and Zoom rooms where these ideas could grow in conversation.

I also want to thank **Jack Dangermond**, founder and president at Esri, whose passion and love for GIS have quietly shaped my entire journey. The way you speak about geography and place—whether on the plenary stage at the Esri User Conference, in informal conversations, or through articles, stories, and initiatives—changed how I viewed, learned, and practiced GIS and all things geospatial. Your words have inspired me, brought me to tears, and reminded me again and again why I chose this path.

You have shown us why this work matters for people, for communities, and for the health of this planet. Thank you for loving this field so fiercely, for protecting it, and for inviting the rest of us to care for it with the same devotion.

Finally, to **you**, the reader. Whether you are a GIS manager, an analyst wondering if leadership might be for you, an executive trying to understand the value of this "map thing," or someone who simply cares about better decisions, thank you for giving this book your time and attention. You are the reason any of this matters. My hope is that somewhere in these pages you find a sentence, a story, or a tool that helps you see yourself more clearly and lead with a little more courage, clarity, and love.

If this book has helped you in any way, know that it is because a whole community stood behind me while I wrote it, for which I am deeply, humbly grateful.

Big love,
—Izabela

Bonus #1: The Executive Influence Toolkit: Speak the Language of Leadership and Get Your GIS Vision Funded.

YOU KNOW YOUR GIS work could transform the organization. You see the inefficiencies, the missed opportunities, the decisions being made without the right information.

But when you try to explain this to executives, something gets lost in translation.

They nod politely. They say, "interesting." Then they move on to the next agenda item, and your budget request sits in limbo for another year.

Here's the truth: It's not that they don't care; it's that you're speaking different languages.

Executives don't think in layers, projections, or spatial analysis. They think in:

- Risk (What could go wrong if we don't do this?)
- Revenue (How does this help us grow or save money?)
- Efficiency (Will this make us faster or better?)
- Compliance (Does this keep us out of trouble?)

This toolkit gives you two things:

1. The 3-Slide Framework—A proven structure for presenting your GIS vision in executive language.
2. Translation Examples—Real scenarios showing how to reframe GIS work as business outcomes.

By the end of this toolkit, you'll be able to walk into any executive meeting and speak their language, not because you're "dumbing down" your work, but because you're finally connecting it to what they're already concerned about.

Let's begin.

The 3-Slide Framework: Your Executive Presentation Structure.

Most GIS presentations fail because they start with the solution ("We need ArcGIS Enterprise!") instead of the problem executives actually care about.

The 3-Slide Framework flips this:

- Slide 1: The Problem (Current State)
    - What's broken, inefficient, or risky right now?
- Slide 2: The Solution (What GIS Enables)
    - How does this change decisions, save time, or reduce risk?
- Slide 3: The Ask (Investment & Timeline)

- ○ What do you need, and what will they see in 90 days?

Why This Works:

- Slide 1 - Establishes credibility: you're not pitching technology—you're naming a pain point they already feel.
- Slide 2 - Connects GIS to outcomes they care about: you're translating "spatial analysis" into "faster permit approvals" or "reduced infrastructure risk."
- Slide 3 - Makes it concrete: you're not asking for faith—you're offering a clear investment with visible milestones.

The rule: If you can't explain your GIS vision in three slides, you don't understand it well enough yet.

Let's build your slides.

Slide 1: The Problem (Current State)

Your goal: Make executives feel the pain of the current situation.

Framework:

"Right now, [*department/organization*] is experiencing [*specific problem*], which is causing [*business impact*]."

Example 1: Utilities

Problem Statement: "Right now, our field crews are spending 3+ hours per day searching for asset information across 7 different systems, costing us approximately

$250K annually in wasted labor and delaying emergency response times."

Why it works:

- Specific numbers ($250K, 3 hours, 7 systems);
- Business impact (cost + safety risk);
- No GIS jargon.

Example 2: Local Government

Problem Statement: "Right now, our permit review process takes 45+ days because staff manually check zoning, flood zones, and utility conflicts across disconnected systems. This delays development projects and frustrates applicants who are comparing us to faster jurisdictions."

Why it works:

- Concrete timeline (45 days);
- Competitive pressure (other jurisdictions);
- Customer impact (frustrated applicants).

<u>**YOUR TURN:**</u> Draft Your Slide 1

Right now, [department/organization] is experiencing ________________________, which is causing [business impact] ___________________________ (specific metrics: time, cost, risk) ____________________.

Slide 2: The Solution (What GIS Enables).

Your goal: Show how GIS changes the game—in their language.

Framework:

"By implementing [*GIS capability*], we will [*business outcome*], which means [*what executives care about*]."

Example 1: Utilities

Solution Statement: "By implementing a centralized GIS asset management system, field crews will access all asset information in one mobile app, reducing search time from 3 hours to 15 minutes per day. This saves $200K annually and cuts emergency response time by 40%."

Why it works:

- Before/After comparison (3 hours → 15 minutes);
- Dollar savings ($200K);
- Safety improvement (40% faster response).

Example 2: Local Government

Solution Statement: "By integrating GIS into our permit review workflow, staff will automatically check zoning, environmental constraints, and utilities in under 2 minutes instead of 2 days. This cuts permit review time from 45 days to 21 days, making us competitive with neighboring jurisdictions and improving our business-friendly reputation."

Why it works:

- Dramatic time savings (2 days → 2 minutes);
- Competitive advantage (faster than neighboring jurisdictions);

- Reputation/economic development angle.

<u>**YOUR TURN:**</u> Draft Your Slide 2

By implementing [GIS capability] ________________, we will [business outcome] ____________ which means [what executives care about] ________________.

Slide 3: The Ask (Investment & Timeline).

Your goal: Make the investment feel concrete, reasonable, and low-risk.

Framework:

"To make this happen, we need [*investment*] over [*timeline*]. In the first 90 days, you'll see [*visible milestone*]."

Example 1: Utilities

The Ask: "To make this happen, we need:

- $150K for GIS software and mobile devices;
- 1 additional GIS analyst (starting Year 2);
- 6 months for full implementation.

In the first 90 days, you'll see:

- Pilot with two field crews using the mobile app;
- Documented time savings from the pilot;
- Rollout plan for remaining crews.

ROI: Full payback in 9 months through labor savings."

Why it works:

- Phased approach (pilot first, then scale);
- Quick, visible win (90 days);
- Clear ROI timeline.

Example 2: Local Government

The Ask: "To make this happen, we need:

- $75K for GIS integration with permit system;
- 4 months for development and testing;
- Training for 12 permit review staff.

In the first 90 days, you'll see:

- Prototype with 3 permit types;
- Measured time savings per permit;
- Staff feedback and refinement plan.

Impact: Permit times drop from 45 to 21 days within 6 months."

Why it works:

- Reasonable timeline (4 months);
- Prototype reduces risk;
- Specific outcome (21-day permits).

<u>**YOUR TURN:**</u> Draft Your Slide 3

To make this happen, we need [Investment] $ ______, [timeline] ________ months, and [resources] ______. In the first 90 days, you'll see ________________________.

The Translation Guide: Stop Speaking GIS, Start Speaking Business.

Executives don't need to understand *how* GIS works. They need to understand why it matters to their goals.

WHAT YOU SAY (GIS LANGUAGE)	WHAT THEY HEAR	WHAT TO SAY INSTEAD (EXECUTIVE LANGUAGE)
"We need to improve our spatial data quality."	"Tech stuff I don't understand."	"Inaccurate data is costing us $X in rework and delays."
"We should implement ArcGIS Enterprise."	"Expensive software request."	"This will reduce permit review time by 50% and improve customer satisfaction."
"Our metadata is incomplete."	"Not my problem."	"Staff waste 5 hours/ week searching for data they can't find or trust."
"We need better geocoding."	"What's geocoding?"	"Emergency responders are delayed because addresses don't match our maps."
"We should create a web map."	"Nice to have."	"Field crews will save 2 hours/day by accessing this information on their phones."
"Our projections are inconsistent."	"Too technical."	"Data doesn't line up between departments, causing errors in planning decisions."

"We need a GIS governance committee."	"More meetings."	"Right now, four departments are buying duplicate data. Coordination will save $50K/year."

The Formula:

Instead of: "We need [GIS thing]."

Say: "[Current problem] is costing us [time/money/risk]. [GIS solution] will [business outcome] in [timeline]."

Real Scenarios: Before & After.

Scenario 1: Requesting Budget for GIS Software Upgrade

Before (GIS Language):

> *"We need to upgrade to the latest version of ArcGIS Enterprise because our current version is three releases behind, and we're missing important functionality like vector tiles and real-time data streaming."*

What executives hear: "We want the shiny new version."

After (Executive Language):

> *"Our public-facing maps are slow and crash during emergencies when traffic spikes. Upgrading our GIS platform will make maps load 5x faster and handle 10x more users, which means residents get critical information during floods and power outages without our system failing. Investment: $40K. Timeline: 90 days."*

What executives hear: "This prevents a public safety and PR crisis."

Scenario 2: Requesting a New GIS Position

Before (GIS Language):

> *"We need another GIS analyst because our current team is overwhelmed with map requests, and we can't keep up with application development and data maintenance."*

What executives hear: "You're busy. Everyone's busy."

After (Executive Language):

> *"Right now, department heads wait 2-3 weeks for basic maps because our team is buried in requests. This delays decisions on $2M+ projects. Adding one analyst will cut response time from 2 weeks to 2 days, which means faster project approvals and better use of executive time. Cost: $85K/year. ROI: Faster decisions on projects worth 20x the investment."*

What executives hear: "This removes a bottleneck that's costing us real money."

Scenario 3: Proposing a GIS Governance Structure

Before (GIS Language):

> *"We should establish a GIS steering committee with representatives from each department to coordinate data standards and prioritize enterprise initiatives."*

What executives hear: "Another committee that will schedule meetings I don't have time for."

After (Executive Language):

> *"Right now, three departments are independently buying the same aerial imagery for $30K each. We're also fixing the same address errors in four different databases. A quarterly coordination meeting with department leads will eliminate duplicate spending and reduce data errors by 60%. Time commitment: 1 hour per quarter. Savings: $50K+ annually."*

What executives hear: "This saves money and reduces redundant work with minimal time investment."

Your Turn: Translate Your Request

Think about a GIS initiative you've been trying to get approved. Let's translate it into executive language.

Step 1: What You've Been Saying (GIS Language)

Write what you've been saying or would naturally say:

____

Step 2: What's the Real Business Problem?

What pain point does this solve? (Think: wasted time, unnecessary cost, compliance risk, slow decisions, safety issues.) _______________________________________

Step 3: What's the Measurable Impact?

Fill in the blanks:

This will reduce/improve/eliminate _______________

____

by [number] ______________________________

which will save/generate/prevent ______________________

____

Step 4: Your Translated Statement (Executive Language)

Now combine it:

> *"Right now, [problem] is costing us [impact]. By implementing [GIS solution], we will achieve [outcome] in [timeline], which means [what executives care about]."*
>
> ____

Test Your Translation.

Before You Present: The Pre-Meeting Checklist.

You've built your three slides. You've translated your request into executive language. Now, let's make sure you're ready for the room.

Pre-Meeting Checklist:

Know Your Audience:

- Who will be in the room?
- What are their current priorities (budget cuts, growth, compliance, public perception, etc.)?
- What have they funded (or rejected) recently?
- Who is your champion in the room?

Anticipate Objections:

- "This sounds expensive." → Your response: "Here's the ROI timeline..."
- "Can't we do this with what we have?" → Your response: "We tried that, and here's what happened..."
- "Why now?" → Your response: "Here's the cost of waiting..."
- "How do we know this will work?" → Your response: "Here's our 90-day pilot plan..."

Prepare Your Evidence:

- Do you have a specific example of the current problem (a project that was delayed, a cost that was incurred, a complaint that was filed)?
- Do you have a comparison (what other jurisdictions/companies are doing, industry benchmarks)?
- Do you have a quick win story (a small pilot or proof-of-concept that worked)?
- Practice Your Ask:
- Can you explain your request in two minutes or less?
- Can you answer "What happens in the first 90 days?" without hesitation?
- Have you practiced with a colleague who will ask hard questions?

The Day Before:

Send a one-page summary of your three slides to attendees. This primes them and gives your champions a chance to support you before the meeting.

Subject line: "Quick preview: [Your Initiative]—90-day plan & ROI"

In the Meeting:

Start with Slide 1 (the problem). Don't lead with your solution. Make them feel the pain first.

Use their language. If they say "operational efficiency," use that phrase. If they care about "community trust," anchor your solution to that.

End with the 90-day milestone. Executives trust incremental progress more than big promises.

If they say no: Ask, "What would need to be true for this to be a yes?" Then listen—you're gathering intelligence for your next attempt.

You're Ready. Now What?

You have the framework. You have the translation guide. You have your 3-slide structure.

Here's what to do next:

This Week:

- Draft your 3 slides using the templates in this toolkit.

- Translate one GIS request into executive language using the formula on page 6.
- Share your draft with a trusted colleague or mentor and ask: "Does this make sense to someone outside GIS?"

This Month:

- Schedule a practice run with a peer or your manager.
- Refine your request based on feedback.
- Request 15 minutes on an executive agenda to present your 3-slide plan.

This Quarter:

- Present your plan and document what questions come up.
- Deliver your 90-day milestone (even if it's a small pilot).
- Report back with results in executive language: "We projected we'd reduce X by Y. Here's what actually happened."

Remember:

- You're not "dumbing down" your work. You're translating your expertise into language that helps leaders make better decisions.
- You're not asking for permission to do GIS. You're inviting them into a partnership where GIS becomes a strategic tool, not a support function.

- You're not just requesting a budget. You're offering a solution to a problem they're already worried about.

You've got this.

Bonus #2: The 90-day GIS Leadership Roadmap Builder: From Chaos to Scarcity: Your First Quarter as a Strategic Leader.

YOU JUST CAME BACK from a conference, and your head is buzzing with ideas. You can finally see how GIS could have a seat at the decision table, how data could flow cleanly through your organization, and how your team could stop living in a permanent "almost caught up" mode.

You can imagine the dashboards executives actually check, the maps field crews actually use, and the meetings where GIS isn't an afterthought.

And then Monday happens.

Your inbox is full before you finish your first coffee. A system patch broke something. Someone needs a "quick" map for tonight's council meeting. A department head forwards a thread: "Can you sort this out?"

By Wednesday, that beautiful vision is buried under other people's urgent requests.

Here's the truth: It's not that you stopped caring, but that you never built a bridge between what you could see and what you could realistically achieve.

A 90-day roadmap is that bridge.

It's not another report to write and forget. It's the simple, practical way you answer one question:

> *"Given the reality of my world, what progress am I actually going to make in the next three months?"*

This workbook will help you:

- Get honest about what you're really carrying right now.
- Choose the 2-3 things that will actually change your trajectory.
- Protect your focus time so the roadmap doesn't die in your inbox.
- Communicate your plan in a way that earns trust instead of skepticism.

By the end, you'll have a one-page roadmap you can share with your team and executives, and actually follow.

Let's begin.

The Dangerous Monday Reality Check.

Before you plan the next 90 days, let's get honest about where you are right now.

Part 1: What Are You Really Carrying?

Take 5 minutes and brain-dump everything on your plate. Don't organize it yet, just list it.

- Big Projects (things with deadlines and stakeholders):
- Critical Maintenance (things that cannot break):
- Ideas You're Excited About (but haven't started):
- The Noise (one-offs, favors, "quick things"):

Part 2: The Honesty Questions:

Now look at your list and answer these:

- How many of these did YOU choose versus how many were chosen FOR you?
- Which of these moves you toward your North Star?
- Which of these are you doing because you're afraid to say no?
- If you could only finish THREE things in the next 90 days, which would change your trajectory the most?

This is your starting point. Not the fantasy version. The real one.

The Three Questions Framework.

Your brain already has a roadmap, but it's running in your head under stress with no boundaries.

A written 90-day roadmap is your decision to stop that.

Here's how to build it: Answer three questions.

Question 1: What Will We Move?

If we could only finish 2-3 meaningful things in the next 90 days, which would change our trajectory the most?

These should be:

- Visible (people will notice when it's done).
- Valuable (it solves a real pain point or unlocks something new).
- Achievable (you can realistically finish it in 90 days, given your capacity).

Your Top 2-3 Priorities for the Next 90 Days:

Priority #1:

- Why does this matter?
- What does "done" look like?

Priority #2:

- Why does this matter?
- What does "done" look like?

Priority #3 (optional):

- Why does this matter?
- What does "done" look like?

Question 2: What Must Stay Stable?

What do we absolutely have to keep running while we work on these priorities?

These are the nonnegotiables:

- The public-facing app everyone uses.
- The nightly processes that feed critical systems.
- The base data that underpins everything else.

If these fail, nothing else matters.

What We Must Keep Stable:

- ______________________________
- ______________________________
- ______________________________

Question 3: What Will We Say "Not Now" To?

What good things will we have to defer to protect our focus?

This is the hardest question, but also the most important.

If you refuse to say "not now," your roadmap becomes a fantasy, and your real plan reverts to "whatever comes in by email."

What We're Choosing to Defer (For Now):

- ______________________________
- ______________________________
- ______________________________

Remember: "Not now" doesn't mean "never," but rather, "not in the next 90 days while we focus on what matters most."

Your One-Page 90-Day Roadmap.

This is the page you'll share with your team and executives. Keep it simple. Keep it honest.

[Your Department/Team Name] - 90-Day Roadmap

Quarter: ______________ 20____

Prepared by: ________________________________

Date: ________________________________

Our North Star (Where We're Heading):

In one sentence, what is GIS for in your organization?

____

What We're Moving (Top 2 to 3 Priorities):

PRIORITY	WHY IT MATTERS	WHAT "DONE" LOOKS LIKE	TARGET DATE
1.			
2.			
3.			

Continue to list the following:

- What We're Keeping Stable (Nonnegotiables):
- What We're Choosing to Defer (Not Now):
- How We'll Protect Focus Time: (Example: "Team will block 2 hours every Tuesday/Thursday for roadmap work. No meetings scheduled during this time.")
- How We'll Know We're On Track: (Example: "Monthly check-in with leadership. Weekly team standup to review progress.")

Special Case: Your First 90 Days in a New Leadership Role.

If you're stepping into a new leadership position (or have just been promoted), your first 90 days will look different.

You need to learn, and you need to show you can lead, at the same time.

The Three Threads of Your First 90 Days:

Thread 1: Listen on Purpose

This isn't passive observation. It's an active investigation.

Questions to ask in your first 30 days:

- How does work really flow into GIS? (Not the official process—the real one.)

- Who are the unofficial influencers? (The people others trust.)
- Where are people quietly doing hero work because they don't trust the system?
- What's the most fragile thing that could break? (And what happens if it does?)
- What's one thing the previous leader did that people loved? (Keep doing it.)
- What's one thing people wish would change? (Consider it.)

Your Listening Plan:

Week 1-2: Schedule 1:1s with:

- Your direct reports: ______________________________
- Key stakeholders: ________________________________
- End users/customers: _____________________________

Week 3-4: Observation:

- Attend team meetings.
- Watch how requests come in.
- Review the backlog/ticket queue.
- Examine deferred or abandoned items.

Thread 2: Stabilize Fragile Systems

Before making changes, ensure that nothing critical is on the verge of failure.

Ask your team: "What's the one thing that keeps you up at night? What **failure** would cause a crisis?"

What Needs Immediate Stabilization:

- ____________________________________
- ____________________________________
- ____________________________________

Action plan: ______________________________

Thread 3: Choose One Visible Win

Resist the urge to chase five medium wins or one giant moonshot.

Pick one change you can deliver in 90 days that makes life better for real people, showcases what GIS can do, or builds trust in your leadership.

Your One Visible Win:

What: __________________________________

Why this one: ____________________________

Who benefits: ____________________________

Target date: _____________________________

Your First 90-Day Roadmap (New Leader Version):

Days 1-30: Listen & Stabilize	**Days 31-60: Choose & Communicate**	**Days 61-90: Deliver & Reflect**
Complete 1:1s with team and stakeholders.	Pick your one visible win.	Ship the visible win.
Identify and fix the most fragile thing.	Share your 90-day plan with the team and leadership.	Gather feedback.
Observe how work really flows.	Start building the win.	Plan your next 90 days with more confidence.

Protecting Your Focus Time (Or Your Roadmap Will Die).

A roadmap without protected time is just a wish list. Here's the hard truth: If your calendar is filled with other people's urgencies, your roadmap will never happen.

The Time Audit.

Before you commit to a 90-day plan, examine your calendar from the last two weeks.

How much time did you spend on:

- Meetings: _______ hours/week.
- Email/Slack/Teams: _______ hours/week.
- "Urgent" requests: _______ hours/week.
- Focused work on important projects: _______ hours/week.

Now be honest:

How much time do you ACTUALLY have for roadmap work each week?_______ hours/week

If the answer is "zero" or "maybe 2 hours if nothing goes wrong," your roadmap is already in trouble.

The Focus Time Commitment.

To move your roadmap forward, you need at least 4-6 hours per week of protected focus time. This is nonnegotiable. If you don't guard it, no one else will.

My Focus Time Plan:

I will block the following times on my calendar for roadmap work:

- Day/Time: ____________________________ (2 hours)
- Day/Time: ____________________________ (2 hours)
- Day/Time: ____________________________ (2 hours)

During this time, I will:

- Close email.
- Turn off Slack/Teams notifications.
- Decline meeting invitations.
- Put "Focus Time - Do Not Disturb" on my calendar.

I will tell my team:

> *"I'm blocking [X hours] per week to work on our roadmap priorities. Unless it's a true emergency, please hold non-urgent questions until [X time]."*

For Your Team:

If you lead a team, help them protect focus time too.

Team Focus Time Agreement:

> *"As a team, we will protect [X hours] per week for roadmap work. We'll use this time to move our top priorities forward, not catch up on email or attend meetings."*

Our team focus time: ______________________________

Communicating Your Roadmap (So It Doesn't Die in a Drawer).

You've built your roadmap. Now you need to share it in a way that earns trust instead of skepticism.

Who Needs to See Your Roadmap?

- Your Team: They need to know what you're protecting their time for.
- Your Manager/Director: They need to know what you're committing to (and what you're not).
- Key Stakeholders: They need to understand why their pet project might be in the "not now" column.

How to Present Your Roadmap?

- To Your Team:

  Script: "Here's our plan for the next 90 days. We're focusing on [2-3 priorities] because they move us toward [North Star]. We're keeping [critical systems] stable, and we're choosing to defer [other things] so we can actually finish what matters. This doesn't mean those deferred items aren't important, only that we're being honest about our capacity. Questions? Concerns? I want to hear them."

- To Your Manager/Leadership:

  Script: "I want to share our 90-day plan and get your input. Given our current capacity and commitments, we believe we can focus on: [list priorities]. We must keep the following stable: [list nonnegotiables]. We're choosing to defer: [list deferrals]. Does this align with your priorities? If something urgent comes up, we can discuss reprioritizing, but I want us to make those decision together."

- To Stakeholders (When They Ask for Something Not on the Roadmap):

Script: "I hear how important this is to you. Here's our current 90-day plan [show one-page roadmap]. If this needs to happen within that window, we'll have to move something else out. Would you be willing to help me decide what that is?"

What this does:

- Shows respect (you're not saying "no," you're inviting a conversation).
- Makes tradeoffs visible (they see what they're asking you to sacrifice).
- Gives you negotiating power (you're not the bottleneck—capacity is).

The Monthly Check-In.

Your roadmap isn't static. Check in monthly.

Monthly Roadmap Review Questions:

1. Are we on track with our top priorities? Yes / No / Partially

   If not, why? ______________________________

2. Is anything on our "keep stable" list at risk? Yes / No

   If yes, what? ______________________________

3. Have we said "not now" to things that tried to creep in? Yes / No

   If no, what got added? ______________________________

4. Do we need to adjust our plan for the next 30 days? Yes / No

   If yes, what changes? ______________________________

When Things Go Sideways (And They Will).

No roadmap survives contact with reality unchanged.

Systems break. Executives change priorities. Emergencies happen.

The question isn't, "Will my roadmap get disrupted?" It's, "How will I handle disruption without abandoning the plan entirely?"

The Roadmap Disruption Decision Tree.

When something urgent lands on your desk, ask:

1. Is this a true emergency?

True emergency = someone's safety is at risk, a critical system is down, or a legal/compliance deadline.

- If YES: Handle it. Communicate to your team what's being delayed.
- If NO: Move to Question 2.

2. Can this wait 90 days?

- If YES: Add it to the "next quarter" list. Thank them for bringing it up.

    - If NO: Move to Question 3.

3. Is this more important than what's on our roadmap?
    - If YES: Have the tradeoff conversation: "We can do this, but it means [X] gets delayed. Is that okay?"
    - If NO: Politely defer: "This is important, but our current priorities are [X, Y, Z]. Can we revisit this next quarter?"

The "Emergency Audit."

If you find yourself constantly in "emergency mode," something deeper is broken. At the end of each month, review your emergencies.

This month's emergencies:

1. ______________________________
2. ______________________________
3. ______________________________

Now ask:

- How many of these were TRUE emergencies (safety, system down, legal)?
- How many were merely "urgent because someone said so?"
- How many could have been prevented with better systems or communication?

If most of your "emergencies" aren't real emergencies, you don't have a capacity problem but rather a systems problem.

(That's what the next bonus will help you fix.)

Your 90-Day Commitment.

You've done the hard work. You've chosen your priorities. You've protected your time. You've built your roadmap.

Now it's time to commit.

My 90-Day Roadmap Commitment.

For the next 90 days, I commit to:

- Focus on these 2-3 priorities:
- Protect [X] hours per week for roadmap work:
- Say "not now" to:
- Check in with my team/leadership on:
- Celebrate progress, not perfection:

What Success Looks Like in 90 Days:

If this roadmap works, in 90 days:

- My team will feel ________________________________
- Our stakeholders will see __________________________
- I will have moved ________________________________
- We will be one step closer to ______________________

My Accountability Partner:

I will share this roadmap with: ____________________

And check in with them: Weekly / Biweekly / Monthly

Signature: ____________________________________

Date: ____

You've Built Your Roadmap. Now What?

You have your priorities. You have your one-page plan. You have protected your focus time.

Here's what to do next:

- This Week:
  - Share your roadmap with your team and get their input.
  - Block off focus time on your calendar for the next 4 weeks.
  - Discuss your plan with your manager/director.
- This Month:
  - Begin work on your top priority.
  - Practice saying "not now" to at least one request that doesn't fit your plan.
  - Check in with your team: Are we on track? What's getting in the way?

- This Quarter:
    - Deliver at least one of your top priorities.
    - Document what worked and what didn't.
    - Plan your next 90 days with more confidence and clarity.

Remember:

Ninety days is the on-ramp, not the highway. This is your first turn in a longer, intentional journey.

You don't need to see the whole path to take the next step.

You don't need anyone's permission to start leading differently from where you are.

You've got this.

Bonus #3: Build a High-Performing Team Without Burning Out: From Hero Mode to Sustainable Excellence.

YOU'RE TIRED. NOT JUST "need a vacation" tired, but the kind that comes from being the person who holds everything together.

You're the one who stays late when a system breaks, who says yes when everyone else is at capacity, the one people come to when they need it done right.

Being the hero feels good—until it doesn't.

Until you realize:

- Your team can't function without you.
- You're the bottleneck in every decision.
- Your best people are burning out trying to keep up.
- New hires either drown or learn to wait for you to rescue them.

Here's the truth you already know but haven't said out loud:

Heroes don't scale, and heroes burn out.

This workbook is about the shift from hero to builder.

Heroes ask: "How can I fix this?"

Builders ask: "How can we design this so it doesn't keep breaking?"

Heroes measure their day by fires put out.

Builders measure progress by how many fewer fires appear.

This framework will help you:

- Assess your real capacity (not the fantasy version where everyone works 60-hour weeks).
- Identify Hero Mode patterns that are quietly destroying your team.
- Map skills gaps so you know what to develop vs. what to hire.
- Hire for character and capacity, not just technical skills.
- Build a development plan that grows people instead of burning them out.

Whether you're a team of one trying to grow or a team of ten trying to stop the chaos, this framework will help you build something sustainable.

Let's begin.

The Capacity Reality Check.

Before growing your team, you need to know your current capacity—not your desired capacity.

Part 1: Your Current Team Snapshot

Team Size: ______

Total team members (including you): _______

Full-time: ____ Part-time: ____ Contractors: ____

Team Roles:

List each person and their primary role:

- ____
- ____
- ____
- ____
- ____

Part 2: Where Does the Time Actually Go?

For each team member (including yourself), estimate the percentage of time spent on: (in %)

TEAM MEMBER	PROJECT WORK	MAINTENANCE	SUPPORT/ REQUESTS	MEETINGS	ADMIN

<u>Now add it up:</u>

Total team hours per week: ________ hours

Hours spent on strategic/project work: ________ hours (______%)

Hours spent keeping things running: ________ hours (______%)

Hours spent on reactive requests: ________ hours (______%)

The Honesty Questions:

Now look at your list and answer these:

How much of your team's time is spent on work that moves the company forward versus simply keeping the lights on?

- ____
- ____
- ____

If someone went on vacation for two weeks, would the team survive without them?

- ______________________________
- ______________________________
- ______________________________

Who is the single point of failure on your team? (The person who, if they left, everything would fall apart)

- ______________________________
- ______________________________
- ______________________________

How much of your capacity is consumed by rework—fixing things that were done wrong the first time?

- ______________________________
- ______________________________
- ______________________________

This is your baseline. Not the story you tell executives. The real one.

Are You in Hero Mode or Builder Mode?

Hero mode feels productive, necessary, and like leadership.

But it's quietly destroying your team's ability to function without you.

The Hero Mode Diagnostic.

Check all that apply to you or your team:

Signs You're in Hero Mode:

- You (or someone on your team) are the only person who knows how critical systems work.
- People wait for you to make decisions instead of making them themselves.
- You regularly work evenings or weekends to "catch up."
- When you're out, things pile up, or people make mistakes.
- You say "it's faster if I just do it myself" more than once a week.
- New team members take 6+ months to become productive.
- You feel guilty taking time off because you know the team will struggle.
- Your team asks permission for things they should be empowered to decide.
- You're praised for "saving the day" but exhausted by it.
- Documentation is always "on the list" but never gets done.

How many did you check?

- 0-2: You're building systems. Keep going.

- 3-5: You're slipping into Hero Mode. Time to delegate and document.
- 6-8: You're deep in hero mode. Your team is fragile.
- 9-10: You're the single point of failure. If you leave (or burn out), everything breaks.

The Cost of Hero Mode.

Hero mode doesn't just hurt you—it hurts your team.

What Hero Mode creates:

For you:

- Burnout.
- Resentment.
- No time for strategic work.
- Inability to take a vacation.

For your team:

- Learned helplessness ("Just wait for [Name] to fix it").
- Lack of ownership.
- Slow skill development.
- High turnover (good people leave because they can't grow).

For your organization:

- Fragile systems that depend on one person.
- Slow decision-making.
- Risk (what happens if the hero leaves?).

The Shift to Builder Mode.

Builders create conditions for repeatable excellence, even without their direct involvement in every decision.

Signs You're in Builder Mode:

- Your team makes decisions without waiting for you.
- New hires have clear onboarding documentation and a 30- to 60-day ramp-up period.
- You can take a week off without everything falling apart.
- People know where to find answers (documentation, processes, and standards).
- You spend more time coaching than doing.
- Mistakes are learning opportunities, not crises.
- Your team asks, "What do you think?" instead of "What should I do?"
- You measure success by how often you're NOT needed.
- Systems and processes are documented and evolving.
- You have time for strategic thinking.

Your Builder Mode Goal:

In the next 90 days, I will shift from hero to builder by:

1. ______________________________

2. ______________________________

3. ______________________________

4. ______________________________

5. ______________________________

What Skills Does Your Team Actually Need?

You can't strategically grow your team if you don't know what's missing.

This assessment helps you see:

- What skills do you have?
- What skills do you need?
- What to develop versus what to hire.

Part 1: Core GIS Skills Inventory

For each skill, rate your team's current capability:

1 = We don't have this.

2 = We have it, but it's fragile (one person knows it).

3 = We have it, and it's distributed across the team.

4 = We have it, and we're excellent at it.

SKILL	CURRENT LEVEL (1-4)	NEEDED LEVEL (1-4)	GAP?
Technical skills			
Data modeling & design			
Database management			
Python/scripting			
Web development			
Mobile app development			
Cartography & visualization			
Data quality & validation			
Systems & tools			
ArcGIS Enterprise administration			
Cloud infrastructure			
Version control (Git)			
CI/CD & DevOps			
Security & access management			
Strategic skills			
Project management			
Stakeholder communication			
Requirements gathering			
Documentation			
Training & knowledge transfer			
Budget & resource planning			

Part 2: Identify Your Critical Gaps

Look at the skills where Current Level < Needed Level.

Top 3 Skills Gaps:

- Impact if we don't address this:
- Impact if we don't address this:
- Impact if we don't address this:

Part 3: Develop vs. Hire Decision

For each gap, decide: Can we develop this internally, or do we need to hire?

Gap #1: ____

Can we develop this internally? Yes / No

If yes, who can learn this? ______________________

Timeline to proficiency: ________________________

If not, do we need to hire? Yes / No

Gap #2: ____

Can we develop this internally? Yes / No

If yes, who can learn this? ______________________

Timeline to proficiency: ________________________

If not, do we need to hire? Yes / No

Gap #3: ______________________________

Can we develop this internally? Yes / No

If yes, who can learn this? ______________________

Timeline to proficiency: ______________________

If not, do we need to hire? Yes / No

Hiring for Character & Capacity (Not Just Skills).

Technical skills can be taught. Character and capacity cannot.

When you hire for skills alone, you get people who can do the work but might not fit your leadership style.

When you hire for character and capacity, you get people who grow with you.

What Is "Character" in This Context? Character is how someone shows up when things get hard.

The Character Traits That Matter Most:

- Curiosity: Do they ask questions? Do they want to understand why, not just how?
- Ownership: Do they take responsibility, or do they blame systems/others?
- Resilience: How do they handle failure, ambiguity, and change?

- Collaboration: Do they make the team better, or do they operate as a solo player?
- Integrity: Do they do what they say? Do they admit when they don't know something?
- Growth mindset: Do they see challenges as learning opportunities?

What Is "Capacity?" Capacity is someone's ability to handle complexity, ambiguity, and growth.

Signs of High Capacity:

- They can handle multiple priorities without falling apart.
- They adapt readily when plans change.
- They learn quickly from their mistakes.
- They don't need perfect instructions to move forward.
- They can work independently and collaborate well.
- They ask for help before they're overwhelmed.

The Hiring Framework

When you're hiring, assess three dimensions:

1. Technical Skills (Can they do the work?)

   Rating: 1-5 ________

   Evidence: ________________________________

2. Character (How will they show up?)

   Rating: 1-5 ________

   Evidence:

3. Capacity (Can they grow with us?)

   Rating: 1-5 ________

   Evidence:

The Decision Matrix

CANDIDATE	SKILLS	CHARACTER	CAPACITY	HIRE?	NOTES
	/5	/5	/5	Y/N	
	/5	/5	/5	Y/N	
	/5	/5	/5	Y/N	

The Rule:

High skills + Low character/capacity =
Short-term gain, long-term pain

Medium skills + High character/capacity =
Long-term asset

Interview Questions That Reveal Character & Capacity:

- For Curiosity: "Describe a time you didn't understand something at work. What did you do?"
- For Ownership: "Describe a project that didn't go as planned. What was your role?"
- For Resilience: "Describe a time you failed at something important. How did you handle it?"

- For Collaboration: "Describe a time you disagreed with a teammate. How did you resolve the disagreement?"
- For Growth Mindset: "What's something you've learned in the last year that changed how you work?"

Listen for:

- Do they blame others, or do they take responsibility?
- Do they focus on what they learned, or **on** what went wrong?
- Do they talk about "we" or "I?"

The Onboarding & Development Plan.

Part 1: The First 90 Days (New Hire Onboarding)

Week 1: Orientation & Context

Goals:

- Understand the organization and team structure.
- Meet key stakeholders.
- Set up tools and access.

Checklist:

- Introduce to the team and key stakeholders.
- Provide access to systems, tools, and documentation.
- Share the team roadmap and the North Star.
- Assign an onboarding buddy.
- Schedule daily check-ins for the first week.

Week 2-4: Learn the Landscape

Goals:

- Understand how work flows.
- Learn critical systems and processes.
- **Complete the first small project.**

Checklist:

- Shadow team members on key tasks.
- Review documentation and standards.
- Complete a small, low-risk introductory project.
- Ask lots of questions (encourage this!).
- Weekly check-in with manager.

Month 2: Build Confidence

Goals:

- Take on more responsibility.
- Start contributing independently.
- Identify areas for growth.

Checklist:

- Lead a small project end-to-end.
- Present the work to stakeholders.
- Identify one skill to develop.
- Biweekly check-in with the manager.

Month 3: Own & Grow

Goals:

- Own a piece of the roadmap.
- Contribute to team improvements.
- Set development goals.

Checklist:

- Own a roadmap priority.
- Suggest one process improvement.
- Create or update documentation.
- Set 6-month development goals.
- Monthly check-ins with the manager.

Part 2: Ongoing Development (For Existing Team)

Development isn't just for new hires. Your existing team needs growth, too.

The Development Conversation Framework

Have this conversation with each team member quarterly:

- Question 1: What are you learning right now?
- Question 2: What do you want to learn next?
- Question 3: What's getting in the way of your growth?
- Question 4: How can I support you?

Part 3: The Individual Development Plan

For each team member, create a simple development plan:

Team Member: ______________________________

Current Strengths:

- ______________________________________
- ______________________________________
- ______________________________________

Growth Areas (Skills to Develop):

- How we'll develop this: ______________________

Timeline: __________________________________

- How we'll develop this: ______________________

Timeline: __________________________________

Stretch Opportunities (Projects to Grow):

- ______________________________________
- ______________________________________

Support Needed: Y/N

Check-in Cadence: Monthly / Quarterly

The Team Health Check.

Capacity isn't just about skills and headcount. It's also about energy, morale, and sustainability. A team with great skills is still fragile if burned out.

The Team Health Diagnostic

Rate your team on each dimension (1 = struggling, 5 = thriving):

DIMENSION	RATING (1-5)	EVIDENCE / NOTES
Workload Balance- Is the work distributed fairly?		
Clarity- Does everyone know what we're working toward?		
Autonomy- Can people make decisions without waiting for approval?		
Growth- Are people learning and developing?		
Psychological Safety- Can people admit mistakes without fear?		
Recognition- Do people feel seen and valued?		
Energy- Do people have capacity left at the end of the day?		
Trust- Do team members trust each other and leadership?		

Total Score: ________ / 40

32-40: Your team is thriving. Keep doing what you're doing.

24-31: Your team is functional but fragile. Address the lowest-rated areas.

16-23: Your team is struggling. Urgent attention needed.

Below 16: Your team is in crisis. Stop adding work and focus on stabilization.

The Lowest-Rated Areas

Identify your bottom three scores:

Area #1: ______________________________

Why is this low? ________________________

One thing we can do in the next 30 days: __________

Area #2: ______________________________

Why is this low? ________________________

One thing we can do in the next 30 days: __________

Area #3: ______________________________

Why is this low? ________________________

One thing we can do in the next 30 days: __________

The Team Conversation.

Don't do this assessment alone. Share it with your team.

Script:

> *"I want to make sure we're building something sustainable. I've been thinking about our team's well-being, and I'd appreciate your input. Could we spend 15 minutes in our next meeting rating ourselves on these dimensions and discussing areas for improvement?"*

What this does:

- Shows you care about their well-being.
- Surface issues you might not see.
- Builds trust through transparency.
- Creates shared ownership of solutions.

From Capacity to Capability: The Growth Roadmap.

You've assessed your capacity, identified gaps, diagnosed Hero Mode, and considered hiring and development.

Now let's put it all together into a growth roadmap.

Your Team Growth Roadmap (Next 6 to 12 Months)

Current State: ______________________________

Team size: ______________________________

Biggest capacity constraint: ________________________

Biggest skills gap: ________________________________

Hero mode score (from page 3): _______ / 10

Desired State (6 to 12 months from now): ___________

Team size: _______________________________________

What we'll be able to do that we can't do now:

____

How work will feel different: ________________________

The Growth Plan

Quarter 1 (Next 90 Days):

Focus: Stabilize and document.

Actions:

- Document the three most critical processes (so they're not just in someone's head).
- Identify one Hero Mode pattern to break.
- Have development conversations with each team member.
- Address the lowest team health score.

Quarter 2:

Focus: Develop and delegate.

Actions:

- Develop one critical skill gap internally.

- Delegate one thing you've been holding onto.
- Create onboarding documentation for new hires.
- Start hiring process (if needed).

Quarter 3:

Focus: Grow and scale.

Actions:

- Onboard new hire (if applicable).
- Launch one team improvement initiative.
- Measure progress on Hero Mode shift.
- Celebrate wins.

Quarter 4:

Focus: Sustain and refine.

Actions:

- Reassess team health.
- Update skills inventory.
- Plan next year's growth.
- Reflect on what worked and what didn't.

Your Commitment:

I commit to building a team that:

- Can function without me being the hero:

  ____

- Grows people instead of burning them out: ______________________________
- Makes decisions without waiting for permission: ______________________________
- Delivers excellent work sustainably: ______________________________

Signature: ________________________ Date: __________

The Hard Conversations You Might Need to Have.

Conversation 1: "You're in Hero Mode, and It's Hurting the Team"

Who needs to hear this: The person (maybe you) who does everything and won't delegate.

Script:

> *"I've noticed you're carrying a lot—probably too much. You're staying late, you're the go-to person for everything, and I'm worried about two things: your burnout and the team's ability to function without you. I need you to start delegating [specific thing] and documenting [specific process]. I know it feels faster to just do it yourself, but we need to build a team that doesn't depend on one person being the hero. Can we talk about what you can hand off this month?"*

Conversation 2: "This Isn't Working, and We Need to Address It"

Who needs to hear this: The team member who isn't meeting expectations.

Script:

> *"I want to talk about [specific issue]. Here's what I'm seeing: [specific examples]. Here's the impact: [how it affects the team/work]. I need to see [specific change] by [specific date]. I'm here to support you—what do you need from me to make this happen? If we don't see improvement, we'll need to have a different conversation about whether this role is the right fit."*

Conversation 3: "We Need More Capacity, and Here's Why"

Who needs to hear this: Your manager/leadership when you need to hire.

Script:

> *"I want to discuss our team's capacity. Currently, we're spending [X%] of our time just keeping things running, which leaves only [Y%] for strategic work. Because of these capacity constraints, we're unable to: [list specific impacts]. To shift from reactive to strategic, we need [additional headcount / contractor support / process improvements]. This would enable us to: [specific outcomes]. Can we discuss what options are available?"*

Conversation 4: "I Need to Stop Being the Bottleneck"

Who needs to hear this: Your team, when you realize you're the problem.

Script:

> *"I've realized I'm slowing us down by being involved in too many decisions. Starting this week, I'm empowering you to make decisions about [specific areas] without waiting for my approval. Here are the boundaries: [what they can decide, what still needs my input]. I trust you. If you make a mistake, we'll learn from it together. I'd rather you move fast and occasionally be wrong than wait for me and move slowly. Questions?"*

You've Built Your Team Framework. Now What?

You know your real capacity. You've identified Hero Mode patterns. You've mapped your skill gaps. You have a growth roadmap.

Here's what to do next:

<u>This Week:</u>

- Complete the capacity reality check (page 2) with your team.
- Take the Hero Mode Diagnostic (page 3) and share your score with someone you trust.
- Identify your top skills gap and decide: develop or hire?

This Month:

- Have development conversations with each team member.
- Document one critical process that's currently only in someone's head.
- Break one Hero Mode pattern (delegate something you've been holding onto).

This Quarter:

- Implement your team growth roadmap (page 8).
- Run a team health check (page 7) and address the lowest score.
- Celebrate progress—even small wins matter.

Remember:

- You can't build a sustainable team while operating in permanent Hero Mode.
- You can't grow people if you're doing all the work yourself.
- You can't scale excellence if it only lives in one person's head.

You've got this!

Disclaimer

THIS BOOK AND THE accompanying bonuses, tools, and examples are provided for educational and informational purposes only. They are based on my personal experience, professional practice, interviews, and research. While many GIS and IT leaders have achieved meaningful results using similar approaches, **no specific outcomes or results are guaranteed**.

Every organization, team, and individual leader operates within a unique context, with different histories, resources, constraints, and cultures. Your results will depend on factors beyond the scope of this book and bonuses, including (but not limited to) your own decisions, the actions of others, and changing organizational priorities.

Nothing in this book or in the bonuses should be taken as legal, financial, or HR advice. You are responsible for applying your own judgment and, where appropriate, seeking guidance from qualified professionals inside or outside your organization.

By using this book and the related materials, you agree that the author and publisher are not responsible for any decisions you make or any outcomes that result from how you choose to implement (or not implement) the ideas and tools presented here.

References

THESE ARE THE WORKS I explicitly draw from for concepts, quotations, or named frameworks.

GIS leadership & strategy:

- Pardue, James. "What It Takes to Be a GIS Leader." ArcUser, Summer 2019, Esri.
- Harper, Alex. "How to Lead with GIS." ArcNews, Fall 2025, Esri.
- Giroux, Paul. The Slimgim Framework: A Pragmatic Geographic Information Management Framework for Assessing and Improving Enterprise GIS and Workflow Health. Master of Science dissertation, Manchester Metropolitan University, 2014.
- Federal Highway Administration (FHWA). Slimgim-T: GIS Capability Maturity Model for State Departments of Transportation – Case Studies of Select Transportation Agencies. U.S. Department of Transportation, 2018.

Leadership, inner work & models:

- Ellis, Robert. *Coaching From Essence: Create a Thriving Practice Doing Powerful Work With Clients You Love.* Futurosity, 2023.

Communication, conflict & difficult conversations:

- Stone, Douglas; Patton, Bruce; and Heen, Sheila. *Difficult Conversations: How to Discuss What Matters Most.* Viking, 1999.
- Harvard Business Review. Selected articles on difficult conversations and listening (e.g., "How to Have Difficult Conversations When You Don't Like Conflict"; "The Power of Listening in Helping People Change").
- Society for Human Resource Management (SHRM). Articles and guidance on managing difficult conversations with employees (e.g., "Managing Difficult Conversations with Employees").
- Time management & prioritization:
- Eisenhower, Dwight D. Attributed quote: "What is important is seldom urgent and what is urgent is seldom important."

Thank You

Izabela Miller

Made in the USA
Monee, IL
14 September 2026

62974524R00167